Nanomaterials in
Architecture and
Art Conservation

Nanomaterials in Architecture and Art Conservation

edited by

Gerald Ziegenbalg | Miloš Drdácký
Claudia Dietze | Dirk Schuch

PAN STANFORD PUBLISHING

Published by

Pan Stanford Publishing Pte. Ltd.
Penthouse Level, Suntec Tower 3
8 Temasek Boulevard
Singapore 038988

Email: editorial@panstanford.com
Web: www.panstanford.com

British Library Cataloguing-in-Publication Data
A catalogue record for this book is available from the British Library.

Nanomaterials in Architecture and Art Conservation

Copyright © 2018 Pan Stanford Publishing Pte. Ltd.

ISBN 978-981-4800-26-6 (Hardcover)
ISBN 978-0-429-42875-3 (eBook)

Contents

Preface

The significance of nanomaterials and nanotechnologies has been growing for many years, and they are now used in many areas, ranging from chemistry and material science via electronics to medicine. Nanoparticles in particular offer unique properties and applications due to their extremely large surface-to-volume ratio. The use of nanomaterials (particles, nanotubes, nanofibres or nanospheres) in the construction sector has been widely discussed. Interest has focussed on the use of nanosized calcium hydroxide particles in stone, mortar and plaster conservation (including wall paintings) since 2001.

The term 'nanolime' has now become synonymous for materials containing finely dispersed calcium hydroxide particles in different organic solvents, mainly alcohols. Over the past few years, significant progress has been made in understanding the reactions of nanolime, its possible applications as well as opportunities and limits of its use. *Nanomaterials in Architecture and Art Conservation* summarises the current state of the art for nanolime, as well as numerous applications and conservation concepts. The results are mainly based on the 'STONECORE – Stone Conservation for Refurbishment of Buildings' project [1], which aimed to develop and test nanolime dispersions, on their own and in combination with conventional stone consolidants such as silicic acid esters. In 2013, the STONECORE project was shortlisted for the annual EuroNanoForum's Best Project Award along with 10 other projects.

It is generally agreed that the nanorange starts with particles whose size is below 100 nm, whereby the typical nanorange starts at sizes below 10 nm. With sizes between 50 nm and 250 nm, the nanolime particles, which were in the centre of interest for applications in architecture and art conservation, lie between the nano- and the microrange. Nevertheless, they offer many new opportunities for the consolidation and strengthening of stone, mortars and renders, as well as wall paintings.

The STONECORE project, which was realised with the support of the European Union in the FP-7 Framework Program, came to an end in 2011. Since then, all of the partners have continued both fundamental and applied research into the use of nanolime for conservation—funded by their own resources or by national programs. Thus, this book provides a comprehensive overview of recent developments in the use of nanolime for conservation treatments. The partners and the lead contributors in the STONECORE project were:

- Geoservice Geophysical Consultants & Contractors, Greece (Klisthenis Dimitriadis, www.geoservice.gr)
- Strotmann & Partner, Germany (Dr. Ewa Piaszczynski, www.restaurierung-online.de)
- Restauro Sp.z.o.o., Poland (Małgorzata Dobrzynska-Musiela, www.restauro.pl)
- Geotron Elektronik, Germany (Rolf Krompholz, www.geotron.de)
- Industrial Microbiological Services Ltd., United Kingdom, (Peter Askew, www.imsl-uk.com)
- University of Fine Arts Dresden, Germany (Prof. Dr. Christoph Herm, www.hfbk-dresden.de)
- Hellenic Ministry of Culture, Greece (Dr. Demosthenes Giraud, www.culture.gr)
- Institute of Theoretical and Applied Mechanics of the Czech Academy of Sciences, Czech Republic (Prof. Dr. Miloš Drdácký, www.itam.cas.cz)
- Technische Universiteit Delft, the Netherlands, (Claudio Patriarca, www.tudelft.nl)
- Univerzita Pardubice, Faculty of Restoration/Department of Chemical Technology of Restoration, Czech Republic (Karol Bayer, www.upce.cz/fr)
- University of Applied Arts, Vienna, Institute of Art and Technology/Conservation Sciences, Austria (Prof. Dr. Johannes Weber, www.dieangewandte.at)
- IBZ-Salzchemie GmbH & Co. KG, Germany (Prof. Dr. Gerald Ziegenbalg, www.ibz-freiberg.de)

The book summarises fundamental investigations to characterise nanolime dispersions as well as possible consolidation mechanisms, formed solids and their distribution in different substrates. Newly developed methods to characterise the action of conservation materials are highlighted as well as the development of new laboratory testing procedures. This is accompanied by chapters that provide an overview of the characterisation and modelling of historic substrates and the deterioration mechanism of natural stones and historic mortars. Special chapters are dedicated to the determination of the physical properties of historic substrates and the use of microscopy to characterise deterioration patterns as well as consolidation effects. A detailed overview of currently available inorganic conservation materials precedes a discussion of the use of calcium hydroxide nanoparticles for conservation.

Nanomaterials in Architecture and Art Conservation has been written for restorers and scientists with the aim of introducing new materials and new conservation concepts. Thus, practical applications and detailed descriptions of conservation actions on many different objects account for a large part of the book. The use of nanolime as both a single consolidant and its combination with silicic acid esters will be discussed comprehensively.

Although the overall thread of the book is to discuss and summarise the application of calcium hydroxide nanoparticles in conservation, the concept of the book is based on independent chapters, where individual views on nanolime applications have been developed by the respective authors. Unfortunately, this has resulted in a minor repetition of experimental details and application advices.

The authors hope to open up and trigger a broad discussion about possibilities and challenges in the use of nanolime dispersions for the conservation of stone, mortar and plaster as well as for the consolidation of wall paintings and stucco.

Gerald Ziegenbalg
Miloš Drdácký
Claudia Dietze
Dirk Schuch
2018

Acknowledgements

This book is based primarily on the results achieved in the joint European project 'STONECORE – Stone Conservation for Refurbishment of Buildings', which was funded by the European Commission within the FP-7 Framework Program on research and innovation under Grant Agreement No. 213651. Without funding, this work could never have been realized, and all of the authors greatly acknowledge the support. The STONECORE project came to an end back in 2011. The book also summarises results that have been obtained over the past years, mainly by research in the following projects:

- ZIM (project number: KU2548202MU2) 2012–2014 "Entwicklung von neuartigen Materialien zur Stabilisierung und Konservierung von Wandmalerei und Putz, insbesondere für salz- und feuchtehaltige, wie auch unterschiedlich saugende Untergründe"
- 'Europäische territoriale Zusammenarbeit Österreich-Tschechische Republik 2007–2013'; Project NANOLITH (M00264) 'Anwendung von Nanomaterialien zur nachhaltigen Konservierung von historischen Skulptur und Architekturobjekten aus Leithakalk'
- Czech Grant Agency project P105/12/G059 Cumulative time dependent processes in building materials and structures
- Czech Grant Agency GA ČR grant 17-05030S 'The influence of the reaction conditions on the formation and mechanical properties of $CaCO_3$ polymorphs in lime-based building materials'
- SAB-Sächsische Aufbaubank (project number 100308774) 'Erweiterung des Baukastensystems zur Sanierung von Stein, Putz und Mörtel'

A special word of thanks is due to all of the project organisations mentioned above for supporting the work on this publication, which allows the results to be exploited by a wider public. The editors and authors thank all of the scientists, graduates and coworkers in the STONECORE project who have contributed greatly to the success of the project, but who are not represented in this book by their own articles, especially Klisthenis Dimitriadis (Geoservice, GR), Rolf Krompholz (Geotron Elektronik, DE), Peter Askew, Jamie Mendoza (IMSL Ltd., UK), Thomas Schmidt (University of Fine Arts Dresden, DE), Demosthenes Giraud (Hellenic Ministry of Culture, GR), Claire Moreau, Dita Frankeová, Ivana Frolíková, Jaroslav Lesák, Jaroslav Hodrment, Pavel Zíma, Miloš Černý, Ondřej Vála, Vladimír Novák, Jan Bryscejn, Libor Nosál (ITAM ASCR, CZ), Elisabeth Mascha (University of Applied Arts, Vienna, AT)Claudio Patriarca, Evert Slob (TU Delft, the Netherlands), Blanka Kolinkeová, Jana Dunajská, Renata Tišlová, Jakub Ďoubal (UPFR, CZ), Jacqueline Pianski and Kathrin Brümmer (IBZ-Salzchemie GmbH & Co. KG, DE).

The late Professor Heinz Leitner, former Head of Conservation of the Wall Painting Department at the University of Fine Arts of Dresden, deserves a particular mention for supporting the idea behind the STONECORE project and organising the project consortium basis. Unfortunately, one of the world's best wall painting restorers and internationally renowned professors died before the project work started. He was only 54. The book is dedicated to his memory.

Chapter 1

Introduction

Gerald Ziegenbalg

IBZ-Salzchemie GmbH & Co. KG, Schwarze Kiefern 4, 09633 Halsbrücke, Germany
gerald.ziegenbalg@ibz-freiberg.de

The conservation of historic monuments is always a complex task that requires an understanding of the historic context of the monuments, the materials and technologies used as well as the opportunities and limits of the conservation materials and techniques that are available. The Venice Charta (1964), which is the most important international framework for the conservation and restoration of historic buildings, contains the following articles, amongst others:

- 'A monument is inseparable from the history to which it bear witness and from the setting in which it occurs.' (Art. 7)
- 'The process of restoration is a highly specialised operation. Its aim is to preserve and reveal the aesthetic and historic value of a monument and is based on respect for original material and authentic documents. It must stop at the point where conjecture begins, and in this case moreover any extra work which is indispensable must be distinct from the architectural composition and must bear a contemporary stamp.' (Art. 9)

Nanomaterials in Architecture and Art Conservation
Gerald Ziegenbalg, Miloš Drdácký, Claudia Dietze, and Dirk Schuch
Copyright © 2018 Pan Stanford Publishing Pte. Ltd.
ISBN 978-981-4800-26-6 (Hardback), 978-0-429-42875-3 (eBook)
www.panstanford.com

- 'Where traditional techniques prove inadequate, the consolidation of a monument can be achieved by the use of any modern technique for conservation and construction, the efficacy of which has been shown by scientific data and proved by experience.' (Art. 10)

The field of heritage conservation and restoration relies on a few essential principles, which new technology must always respect:

- 'Primum non nocere, secundum cavere, tertium sanare' (first do not harm, second be careful, third heal)
- As non-invasive as possible
- Comprehensive documentation before, during and after treatment
- Reversibility/retreatability

In this context, one of the most widely discussed issues concerns the compatibility of the materials used for conservation, which should cover, for example:

- Mechanical properties
- Physico-chemical properties
- Chemical/mineralogical composition
- Weathering behaviour
- Appearance

Most conservation projects deal with objects that have already undergone several modifications, restorations and refurbishments in their lifetime so that the aspect of the compatibility of different materials becomes particularly important. It is a well-known fact that the conservation of natural stone by cementitious materials has often caused significant, additional damage because certain fundamental properties of the different materials did not match. Similar problems have arisen with the use of organic polymers. In other words, compatibility is required to:

- The original construction components
- All constituents of a construction or monument
- The components used in any former restoration or refurbishment steps

The conservation of inorganic materials such as stone, mortar or plaster requires traditional inorganic consolidants. A rough classification allows a distinction between:

- Typical lime-based materials such as slaked lime, white lime hydrate or dispersed white lime hydrate
- Natural hydraulic lime mortars, injection grouts or plasters
- Gypsum-based systems
- Cementitious systems, including Roman cement and various pozzolanic-cement mixture systems

Traditional organic materials for stone conservation include various waxes, oils, acrylates and epoxides.

One of the most successfully used conservation material that allows the consolidation of heavily disintegrated structures are silicic acid esters. These lead to the formation of silica gels, which stabilise deteriorated zones. The main advantage of silicic acid esters are that the solutions can penetrate deep into deteriorated structures without the drawbacks of lime hydrate suspensions. Silicic acid esters allow the structural consolidation of sandstone and related materials, but they are unable to bridge voids or fissures. Comparable materials that allow the structural consolidation of calcareous stones by the formation of carbonatic structures were not known before 2008. This was the starting point of the European project 'STONECORE – Stone Conservation for Refurbishment of Buildings', which was financed within the FP-7 Framework Programme of the European Union [1]. The main objectives of the project were:

- The development of nanomaterials compatible to natural and artificial stone for refurbishing buildings, monuments, fresco, plaster and mortar
- The development of nanomaterials suitable for the safe and environment-friendly removal of mildew and algae
- The development and testing of suitable technologies for the application of nanomaterials in refurbishment
- The development and testing of non-invasive or minimally invasive assessment methods

The STONECORE project, in which all of the authors of this book were involved, came to a successful end in 2011. Unique sols containing nanoparticles of $Ca(OH)_2$ stably dispersed in different alcohols were developed and successfully applied on several objects. Their properties and applications are the focus of this book. It presents and discusses fundamental investigations to characterise nanolime dispersions as well as experiences resulting from comprehensive laboratory tests of differently composed nanolime dispersions in detail.

Nanolime allows consolidation by carbonation, similar to the processes that occur during the setting of conventional lime mortars, though also in combination with silicic acid esters. Possible carbonation and consolidation mechanisms, formed solids and their distribution in different substrates will be highlighted. This is accompanied by chapters that provide an overview of the characterisation and modelling of historic substrates and a summary of deterioration mechanisms for natural stones and historic mortars. Special chapters are dedicated to the determination of physical properties of historic substrates and the use of microscopy to characterise deterioration patterns as well as consolidation effects.

Before discussing the use of calcium hydroxide nanoparticles for conservation, a detailed overview of currently available inorganic conservation materials is provided.

The book has been written for restorers and aims to introduce new materials and new conservation concepts. Consequently, practical applications and detailed descriptions of conservation actions on different objects account for a large part of the book. The use of nanolime as both a single consolidant and its combination with silicic acid esters will thus be discussed at length.

Since the end of the STONECORE project, many papers have been published on the use of nanolime in conservation. The most important ones will be summarised and discussed. The authors hope to open up and trigger a broad discussion about possibilities and challenges in the use of nanolime dispersions for conservation of stone, mortar and plaster as well as for the consolidation of wall paintings and stuccos.

Chapter 2

Historic Substrate Characterisation and Modelling

Johannes Weber,[a] Miloš Drdácký,[b] and Gerald Ziegenbalg[c]

[a]*University of Applied Arts Vienna, Institute of Art and Technology/Conservation Sciences, Salzgries 14/1, 1010 Vienna, Austria*
[b]*Institute of Theoretical and Applied Mechanics of the Czech Academy of Sciences, Prosecká 76, Prague 9, Czech Republic*
[c]*IBZ-Salzchemie GmbH & Co. KG, Schwarze Kiefern 4, 09633 Halsbrücke, Germany*
johannes.weber@uni-ak.ac.at; drdacky@itam.cas.cz; gerald.ziegenbalg@ibz-freiberg.de

2.1 Deterioration of Stone and Historic Mortars*

2.1.1 Natural Stone: Introduction

Natural stones are assemblages of one or several types of minerals. Unlike solids like metals, glass or synthetics, stones are non-uniform and sometimes also anisotropic systems. Just on a macroscale of observation, they are often considered uniform, a simplification on which a majority of the test procedures related to mechanical properties of stones are based.

The factual non-uniformity or heterogeneity of stones is determined by their microstructural composition of mineral grains and pores of varying type, size, shape and orientation, each of the

---

*This section is written by Johannes Weber.

Nanomaterials in Architecture and Art Conservation
Gerald Ziegenbalg, Miloš Drdácký, Claudia Dietze, and Dirk Schuch
Copyright © 2018 Pan Stanford Publishing Pte. Ltd.
ISBN 978-981-4800-26-6 (Hardback), 978-0-429-42875-3 (eBook)
www.panstanford.com

crystalline minerals forming an isotropic body itself. When mineral grains and pores are arranged in a random way, we speak about an anisotropic stone with no preferred orientation in respect to its macroproperties. Yet, the link between these properties and the features of the microstructured is complex to establish, a reason why the understanding of material behaviour and failure differs so much between the engineering viewpoint and the perspective of a microscopist. This applies even more for anisotropic textures with layering, bedding, schistosity or flow textures, respectively, which appear frequently in many rocks of sedimentary, metamorphic and eventually magmatic origin. Measurements of macroproperties of such stones have to take into account that most parameters would significantly differ in dependence of the orientation.

In general, the boundaries between mineral grains usually represent the weakest places within a stone fabric, in particular when they are formed in a way to allow for empty spaces. Such stones which possess a significant amount of voids are called porous, and in frequent cases where the pores are interconnected to form an open-pore system, their fabric is accessible to liquids and gaseous molecules. Non-porous stones are called compact, though even these may have small spaces along grain boundaries in the submicron range.

2.1.2 Classification of Rocks

The usual criteria to classify rocks in geological science are based on their origin, resulting in the following three large groups:

Igneous rocks are formed by cooling from liquid magma. Commonly of siliceous nature, they are subdivided into volcanic and plutonic rocks. Formed by rapid cooling at the surface of the earth, volcanic rocks are characterised by a fine-grained, often even vitreous matrix in which larger crystals are embedded. A frequent representative of this group is basalt. Plutonic rocks, on the other hand, have been formed at much lower cooling rates in some depth of the crust where the rising magma became stuck. Consequently, they differ from the former group by their roughly equicrystalline size of minerals. By orogenetic events, plutonic rocks are eventually lifted to the surface where they can be quarried. Well-known representatives are granite, diorite and gabbro. Out of these, only granites are sufficiently rich in silica to form quartz, an important aspect for stone working, since this mineral is the hardest.

Metamorphic rocks originate from other rock types by transformation at more or less high temperature and/or pressure when transferred to deeper zones of the earth's crust by orogenetic processes. Depending on the type of pre-existing rock, they can be of siliceous or of carbonate nature. The alterations caused by metamorphism differ in type and degree in function of its intensity and several other parameters. Commonly, the pre-existing petrographic structure is changed profoundly. The porosity is diminished due to compaction, crystals tend to recrystallise to larger sizes and new minerals are formed by reaction from certain assemblages which become unstable. Frequently, directed tectonic stresses lead to oriented crystal growth and parallel texturing. When these stresses outlast crystallisation, this can result in schistous textures with poor mechanical properties. Hence, there is no common rule of whether metamorphism would improve the strength of a rock, but in general it appears that soft rocks become stronger and strong ones often weaker during their metamorphic transformation. Out of a large number of metamorphic lithotypes, marble is probably the one with the widest use in sculpture and architecture. Having originated from limestone, marble is composed of recrystallised carbonate minerals.

Sedimentary rocks are formed from sediments by diagenetic processes under various conditions. According to their origin, they are classified into several different groups, out of which detrital or clastic sediments are the most common representatives in sculpture and architecture. Clastic sediments are formed from fragments of pre-existing rocks by the action of weathering agents, transport by water or air, and sedimentation. The components vary in composition, though due to its resistance against weathering, quartz is the dominant mineral in most sands. When either carbonate crystals or fine siliceous debris from circulating waters are precipitated or sedimented in pores and at grain boundaries, thus cementing the components together, a solid sedimentary rock forms from a loose sediment. In that way, for example, sandstone forms from sand; if quartz is the prevailing component, such a rock is called quartz sandstone. It is, however, largely the cementing or pore-filling minerals which control the durability of detrital sediment, since those minerals are most easily attacked by water migrating through the pore system. Other than quartz, carbonate minerals such as calcite can be the major species forming the components. One should then avoid using the term 'sandstone' and rather refer to calcareous arenite. Both quartz and calcareous arenites play important roles in the sculpture and architecture of many regions. Another important group of sedimentary carbonate rocks is that of compact limestones. Most of them originate from the chemical precipitation of calcium carbonate, calcite, in a marine environment. Fossil shells may contribute to a large extent to the formation of these rocks, so a combined chemical and biogenic origin may describe best the conditions of their origin. The carbonate crystals formed from the original mudstone are usually of a submicron size, so-called micrites. They assemble to a compact stone with a tendency of cleavage and cracks, which, however, may have healed by secondary carbonate crystals, a feature contributing to the interesting pattern of some brands. Other varieties have their cracks filled with less stable clayey matter. Another factor of value is the colour of a limestone, which can range from yellow over red to grey or black, depending on the nature of colouring impurities such as iron compounds or bituminous substances.

Table 2.1 gives a rough overview of the lithological classes which are mostly used for monumental stone objects.

2.1.3 Weathering of Building Stones

2.1.3.1 General remarks

Most natural building stones are extracted from rocks which had originally formed at conditions much different from the ones at the earth's surface. When these stones are extracted from the quarry, their fabric and mineral assemblages have since long been in a metastable state. Even if they might remain unaltered for another millions of years in a metastable equilibrium, they are principally prone to alteration, especially when exposed to the atmosphere. The process of alteration under the influence of the atmosphere is called weathering. It is for obvious reasons that a stone artefact is

Table 2.1 Lithological classes which are mostly used for monumental stone objects

Minera-logical nature	Type of fabric	Origin	Examples	Remarks
Silicates (quartz, feldspars, clay, etc.)	Compact	Plutonic	Granite, diorite	Quartz only in granite (hard)
		Volcanic	Porphyry	Relatively rare use
	Porous		(Basaltic) Lavastone	Irregular pores, partly closed
	Compact	Metamorphic	Gneiss, serpentine	Often anisotropy of fabric (oriented microstructure)
Carbonates (calcite, dolomite, etc.)	Porous	Sedimentary (clastic)	(Quartz) Sandstone	Often carbonatic pore cement or clay as pore filler
			(Bio) cal-careous arenite, soft lime-stone	Some sorts contain minor amounts of silicates (e.g., clay)
	Compact	Sedimentary (chemical-biogenic)	Limestone	Micritic calcite prevails; stone is sometimes erroneously called 'marble'
		Metamorphic	(White) Marble	Rel. large crystal size; most prominent sculptural stone

usually far more exposed to atmospheric impacts than a rock in its geological context. The alteration of rock outcrops due to weathering is primarily a natural process. It results in the formation of landscape reliefs and soils, thus forming a pre-requisite for life on earth. In dependence on the chemical, mineralogical and petrographic nature of a rock, further as a function of pre-existing structural defects, the relief energy of the surface and the climatic and hydrogeological conditions, type and degree of weathering may vary considerably. In construction and landscape engineering, efforts to stabilise the ground form an important part of many projects, though their results are usually only temporary and limited to small areas, given the enormous power of the processes aimed at bringing the earth surface towards equilibrium.

In the case of stone artefacts as objects of conservation, the level of tolerance towards material losses by weathering processes is of course much lower than in the context of engineering: wearing off a sculpture or architectural surface in the range of even a millimetre may mean to lose much of the artistic or historic value, especially when paints, workman traces or other surface features of value are present. This is why conservation scientists have to be concerned not only about the bulk of a stone object but also about the sensitive surface. From that viewpoint, the assessment of material parameters in the range of millimetres or centimetres, usually from the surface inwards, is of high relevance for any successful measure of conservation.

Similar to the natural weathering of rocks, the deterioration of stone in sculpture and architecture is a complex process, and much of the knowledge achieved in the respective disciplines of geological engineering can be used to understand the causes and mechanisms which seriously threaten our cultural heritage. Yet, the scale is a different one: while natural rock decay affects us in the centimetre to meter range, a stone artefact suffering surface alteration or losses even within the range of a millimetre can be a matter of concern. The intention to preserve a stone object is not limited to its general shape obtained by carving or cutting but also includes the surface as a potential carrier of valuable artistic or historic information. Under these conditions and taking into account the limits to invade a work of art, conservation of heritage objects is facing a number of highly specific problems when assessing and measuring materials in their sound and weathered states and when designing and evaluating measures of conservation.

Conservation scientists frequently base their systematic approach to stone decay on the three possible factors triggering relevant processes, that is, chemical, physical and biological attack. It is worth mentioning that this type of classification, though of value because it is related to the important question of origin of a given defect, is nevertheless of little help in assessing the type of decay as such, probably because these factors interfere with each other in multiple ways and may result in different patterns of defect. A phenomenological approach is therefore more useful when direct interventions are planned to conserve the material, while preventive measures have to account for the cause of decay in the context of the object and its environment.

2.1.3.2 Weathering phenomena and processes

The illustrated glossary on stone deterioration patterns, published in 2008 by the International Scientific Committee for Stone Conservation, ICOMOS [2], is an excellent tool to describe, by means of an agreed terminology, the usual alteration phenomena found on the surface of natural stone artefacts. The main categories, each one containing a number of detailed decay patterns, are as follows:

- Crack and deformation (fracture – star crack – hair crack – craquele – splitting)
- Detachment (blistering – bursting – delamination – disintegration – fragmentation – peeling – scaling)
- Features induced by material loss (alveolisation – erosion – mechanical damage)
- Discoloration and deposit (crust – discolouration – efflorescence – encrustation)
- Biological colonisation

The usual structural defects leading to the loss of material and hence likely in need of consolidation are to be found within the two categories Crack and Detachment, which encompass the most typical phenomena found on deteriorating stone objects. Three of the most frequent types of stone decay can be identified within these phenomenological classes, all three of them defining failures in need of consolidation. They are:

- Surface parallel failures: 'delamination'
- Loss of cohesion between grains: 'disintegration'
- Cracks and fissures propagating into the interior of the stone: 'crack and deformation'

Group 3 type of failure, which usually affects compact lithotypes rather than porous ones, comprises a large variety of possible cracks and fissures of varying width and length. The possible mechanisms responsible for their formation, range from lithologic preconditions such as cleavage to specific external factors such as mechanical shock or static problems. It seems therefore impossible to develop a common model for crack formation in the frame of the present contribution, though it is beyond doubt that such defects deserve the highest attention within the conservation of a stone object.

Weathering phenotypes of groups 1 and 2, on the other hand, are usually based on a limited number of stress factors and decay mechanism. They will be discussed more in detail later.

In principle, the question, which of the above phenomenological patterns develops in a specific case depends on several factors such as the general petrographic characteristics of the lithotype, the type and origin of externally induced stresses and steepness of their in-depth gradient from the surface and the existence of layered structures parallel to the surface, be they formed by coatings, conservation treatments or earlier weathering.

In respect to the petrographic characteristics, two major criteria are of particular relevance to the durability of a given lithotype against weathering, namely its mineral composition and the microstructure or fabric. In respect to the former, silicate minerals are chemically more stable compared to carbonates. Even if some of them such as feldspars are known to be prone for chemical alteration and decomposition, this holds for geological timescales rather than for the lifetime of a stone artefact. Amongst the vast group of silicate minerals, quartz is especially resistant, both from a chemical and from a mechanical point of view. Other silicates can be considered as mechanical unstable members of the group: clay minerals tend to swell when water accesses their abundant intracrystalline planes, a mechanism which leads to high rates of swelling of lithotypes containing clay.

Carbonate minerals, on the other hand, are chemically far less stable than their silicate counterparts. The theoretical rates of dissolution of calcite, for example, depending on pH and temperature of the aqueous solution, are in any case an order of magnitude higher than for the common silicate minerals. Dissolution and precipitation of limestone, processes responsible for the formation of whole landscapes in karst areas around the world, are likely acting on limestone artefacts. Under the real conditions of an exposed monument, however, the grain size and the compactness of the grain fabric play an important role in the way that larger calcite crystals suffer a lower rate of dissolution than fine ones, provided that water is able to access them through the pore system of the stone. The accessibility to water explains why the minute crystals of calcite along the margins of pores forming the cement of many detrital sedimentary stone types are so prone to weathering. In addition, the attack of calcite by acids other than carbonic includes the formation of more soluble salts as reaction products. The problem of gypsum, a partly soluble calcium sulphate hydrate formed on or near the surface of calcitic building materials by the action of sulphurous air pollutants, constitutes a major issue in conservation of sculptures and facades, especially in urban areas. Even if the level of SO_2 emissions has recently been drastically reduced, the gypsum formed over more than 100 years of industrialisation remains there and has contributed to the build-up of mechanically fragile profiles at the surfaces of stones and renders. Further on, as mentioned before, the second parameter of importance is the microstructure of a stone, a term which comprises, amongst others, not only size, shape and intergrowth of minerals but also type, size and connectivity of pores. The mechanical strength usually understood and measured as a macroproperty of a given stone is in fact more governed by the microstructure than by the nature of the minerals. Even if the strength of a stone cannot be directly related to its weathering resistance, the microstructure or stone fabric is of equal importance for both of these properties.

Having in mind the principle processes of decay by weathering, it is useful to group the large variety of building stones into a highly simplified system of lithotypes, basing it on their most relevant properties, namely silicate versus carbonate and compact versus porous stones.

2.1.3.2.1 *Delamination*

Surface parallel failures are probably the decay phenomenon most frequently encountered on stones. To understand the principle processes involved, it is worthwhile to recall that most stresses induced to a stone object are predominantly due, or at least governed by, external factors of environmental origin. If in a first approach we consider a stone object as a uniform solid of given properties in respect to its thermal conductivity and moisture migration, then all atmospheric impact of energy and mass will act via the surface towards the interior and result in stress gradients, at least as long the environmental conditions are not constant over time. A simple example is the cycle of moistening by atmospheric precipitation followed by drying. The amount of water present in the pore structure of the stone will strongly differ between subsequent depth zones. Thus, the resulting moisture-related swelling of the stone matrix will differ in its amount, usually decreasing from the moist surface to the dry interior of the object, that is, the surface layer will be subjected to swelling more than the interior layers. In this way, shear stresses between adjacent layers will build up, which, depending on the number of cycles, the moisture gradients and of course on the strength of the stone fabric, may lead to losses at some stage. The importance of an in-depth gradient built up by environmental cycles can be seen by the fact that stone elements constantly soaked with water, for example, in fountains or constructions under water, are usually more durable than those exposed to atmospheric cycles of moistening and drying. In view of these, porous lithotypes are the typical candidates for detachment caused by moisture, especially if their pore size is in the narrow range. First, because they would soak water to the interior, but not very efficiently, so that steep moisture profiles are likely to build up. Second, such stones exhibit often high rates of swelling when wet. Many clayey sandstones belong to this group, while in open porous lithotypes such as most of the calcareous arenites water is soaked more readily; hence gradients of moisture are less steep, and drying occurs more rapidly. This advantage is counterbalanced by the aforementioned chemical sensitivity of such stones when compared to quartz arenites.

Similar to moisture-induced swelling, frost and salt crystallisation are two other factors resulting in volume expansion; just as moisture, they affect fine porous stones where they can cause various forms of delamination. For compact lithotypes with virtually no absorptivity for water, on the other hand, there is just thermal insulation as the external factor prone to create differential dilatation along the depth profile of the stone. In extreme cases, this might lead to delamination, whereas cracking seems to be a more common response of compact stones to thermal stresses. This is probably due to higher thermal conductivities in compact solids as compared to porous ones, by which strong temperature gradients are unlikely to build up and stresses affect the bulk rather than just a surface layer of the element.

Crystalline marble, though belonging to the family of compact stones, is an exception in that it behaves in specific ways towards thermally induced stresses. Especially sensitive to thermal stress due to the anisotropic expansion of calcite, the marble fabric is in fact a compact assemblage of rather coarse crystals whose relatively even grain contacts and intracrystalline cleavage are easily subjected to sliding or decohesion, respectively. The former is responsible for the frequently observed bending of marble slabs under the impact of single-sided sun radiation, while the latter process leads to a slight opening of the marbles crystal structure, sufficient to allow for capillary water absorption followed by chemical, biological or physical attacks which finally result in sugaring rather than delamination. When, however, acid rain is entering the marble structure and not efficiently washed off the surface, repeated cycles of carbonate dissolution, migration and precipitation, for example, of gypsum, would form layers of differing properties. In that case, scaling or other subtypes of detachment would appear, similar to porous limestones, as described next.

2.1.3.2.2 *Granular disintegration*

The term 'granular disintegration' refers to the detachment of single grains, or small groups of grains, on a microscale. As pointed out in the ICOMOS glossary (page 20), 'The grain size of the stone determines the size of the resulting detached material' [2]. According to the size of disintegrating grains, the terms 'crumbling', 'sugaring', 'sanding' and 'powdering; are proposed. With the exception of sugaring of crystalline marble, as discussed earlier, disintegration is nearly always limited to granular sedimentary rocks, which are separated by intergranular pores and diagenetically cemented by secondary minerals such as calcite, authigenetic quartz, clay minerals or iron hydroxides. Granular sedimentary rocks are usually of detrital origin, that is, they are composed of more or less fine fragments formed in the course of weathering of parent rocks followed by transport, sedimentation and finally compaction and cementation, the latter being subsumed under the term 'diagenesis'.

Diagenetic processes usually do not entirely close the interstitial spaces between the detrital grains. Consequently, they are not in contact with each other at the full length of their boundaries, leaving intergranular spaces which are partly empty and hence form the places where moisture may migrate and internal stresses would act. As the mineralogical nature of the components differ between siliceous and carbonate—the former lithotypes are called quartz arenites or sandstones; the latter calcareous arenites—their sensitivity to acid attack differs greatly. Thus, siliceous or quartz sandstones are chemically resistant, and if compact layers are formed in the course of weathering, these are rather due to particulate matter deposited in the sub-surface pores rather than to corrosion products. When scaling is observed on pure siliceous sandstone related to moisture- or frost-driven swelling cycles, or to the accumulation or crystallisation of soluble salts, it is limited to rather fine porous or compact varieties with relatively low water absorption. More frequent is the case of loss of cohesion between the sand grains by stresses induced right from inside the pores, or by chemical dissolution of the cementing material. Sanding is therefore a predominant decay pattern for such sandstones.

On the other hand, deterioration of calcareous arenites is usually connected to chemical processes of dissolution by the pore liquid, which preferentially affects the secondary cementing minerals. Ion transport by capillary migration of the pore solution would then lead to the precipitation of minerals in certain zones of the stone element. In this way, layers affected by dissolution may alternate with layers of precipitation, both strongly differing in their petrophysical properties, so that the combined effect of granular disintegration and layer formation would result in delamination and sanding at the same time. The frequent case of sulphuric acid attack on calcium carbonates is a good example to illustrate this. The resulting reaction product, calcium sulphate in the form of gypsum, is soluble enough to be dissolved by water and transported a few millimetres towards the surface through the pore system but sufficiently insoluble not to leave the material but to crystallise at or beneath the surface during drying by evaporation. In this way, a close sequence of horizons is formed by the partly dissolution of calcite on the one hand and the precipitation of gypsum on the other hand. As a result, two or more layers with significantly different material parameters are forming on top of the sound part of stone, with the compact gypsum layer towards the surface. Where no direct rain or run-off water approaches the stone, the gypsum layer does not get dissolved nor washed out but remains as a topping, likely to hinder vapour diffusion from the interior and most probably prone to hygric or rather hygroscopic swelling by vapour uptake from the atmosphere. The most drastic examples of detaching gypsum and stone layers can be found on small-scaled porous limestone elements such as, for example, pinnacles, exposed at elevated levels of a building to highly differing conditions of moisture input and drying from either side and hence frequently subjected to a directed migration of moisture through the pore system. Such elements clearly reveal decay patterns of granular disintegration on the rain-exposed side and crust formation followed by delamination and finally sanding on the protected side where the layered position of zones of dissolution and precipitation is pronounced.

Compact carbonate stone types reveal different decay patterns, depending, however, on their grain fabrics. The carbonate crystals in compact sedimentary limestones are usually extremely fine and well-interlocked along their boundaries, unless through lithological defects such as clayey layers, veins and fissures, no water can enter the stone efficiently. Yet, its carbonate surface is slightly soluble by atmospheric water, which results in a uniform and slow wearing off in exposed areas. The run-off water is to some extent saturated with gypsum, which tends to precipitate on adjacent surfaces of the object, leading to the formation of black crusts on a hardly decayed stone. Thermal stresses due to insolation may occur and they would probably result in the detachment of stone layers if not steady dissolution would outdistance that process.

Similar effects to the one described earlier are, for example, caused by thermal radiation or freezing or by internal factors such as crystallising salts accumulated in the surface layer. All these stresses result in dimensional changes and shear stresses parallel to the surface of the object. The effect can be strongly increased when the object itself reveals layers of different dilatation coefficients, as, for example, due to coatings, surface treatments or profiles of chemical weathering.

All these mechanisms can be subsumed under the term 'physical weathering', even if they are frequently triggered by chemical or biological processes. The effects of alteration are, amongst others, always reflected by changes of the microstructure in the affected zones of the stone. To level these differences out, for example, by removing accumulated products of corrosion or, inversely, by depositing new binder in the pore space of affected layers, is the main target of conservation. When this is not matched in a reasonably good way, that is, when an existing petrophysical profile is not evened out or when it is even enhanced by a treatment, the measures are likely to produce negative effects rather than increasing the lifetime of the stone artefact. Illustration of the formation of layers of different porosity caused by the dissolution related processes is summarised in Table 2.2.

2.1.3.3 Principle conclusions and issues of conservation

The most frequent defects of the stone fabric due to surface weathering have been discussed on a general level. It is evident that almost all forms of weathering result in a relatively steep profile of sometimes significantly different petrophysical properties from the surface of a stone element to its interior. Test and measurement

Table 2.2 Illustration of the formation of layers of different porosity caused by the dissolution-related processes

Picture	Description
	Typical fabric of a **plutonic rock**, for example, granite. The silicate crystals, chemically stable and usually well-interlocked by their size and shape, leave no space for an open pore system. If at all affected by weathering, this would either be limited to the surface of the element or result in the formation of cracks. However, for the ease of workmanship, a great deal of geologically weathered granites have been used for stoneworks in monuments; such granites may be more sensitive to decay than sound ones.
	Typical fabric of marble, a **metamorphic rock** composed of coarse carbonate crystals such as calcite or dolomite. Grain boundaries are more or less even, with a low degree of interlocking. In combination with the pronounced internal cleavage of the carbonate crystals, external stresses may cause sliding movements of the crystal fabric, eventually leading to macrodeformation of the whole element. Alternatively, the grain contacts may open and disintegration—sugaring—may occur.
	Typical fabric of **detrital sediment**, for example, sandstone. Despite the mineralogical nature of the components—quartz or other silicate minerals, or carbonates like calcite or dolomite—the intergranular spaces usually define the weakest places of the fabric. Consisting of pores and various minerals of secondary origin, these spaces are prone to various types of physical, chemical and biological attack, all these stress factors being linked to water entering the pore system.
	Layered texture of a detrital sedimentary rock with bedding planes formed by alternating coarse and fine-grained layers. It is evident that the petrophysical properties of the stone vary significantly between adjacent layers. The orientation of bedding planes towards the surface of the element and in respect to the load are decisive for the forms of weathering to occur on that stone.
	Typical profile formed by chemical weathering, especially found in granular calcareous lithotypes. The zone in the centre has suffered granular disintegration by dissolution of cementing minerals; by capillary migration of the solution towards the surface and precipitation of salts—usually gypsum—in the sub-surface zone and on top in form of a crust (top), a sequence of layers has formed which will result in delamination due to physical stresses. Small elements exposed to precipitation on one side, and protected on the other, are typical candidates for this form of weathering due to continuous or repeated moisture migration towards the protected surface.
	Alternative option to the above figure: Characteristic profiles found in granular lithotypes due to surface weathering. (a) Surface erosion by chemical or mechanical attack without microstructural defects of the interior; (b) granular disintegration at an increasing degree towards the surface; (c) granular disintegration within an interior layer parallel to the surface, resulting in delamination of the surface; (d) compact outer crust largely formed by deposition, with minor contribution from the substrate causing some granular disintegration below the surface.

procedures to account for such profiles are in need of high enough spatial resolution, a fact which explains why most of the existing normative tests are just of limited use in conservation sciences.

The conservation of a stone object in its weathered state forms an important step in the frame of its restoration. In the majority of cases, conservation is linked to measures of consolidation or strengthening of the stone fabric in the altered layers, usually by introducing a liquid capable to deposit a solid inside the pore system either by chemical reaction or by evaporation of the liquid phase. According to each specific phenotype of weathering, this task may be approached either by impregnation via the surface or by other means of application enabling selective consolidation, for example, by injection into the defective layers.

The selection of appropriate products of consolidation must consider several factors, such as its compatibility with the substrate from a chemical viewpoint and its capacity to bridge defects inside the stone fabric to a sufficient degree, to name just a few criteria. The primary requirement, however, relates to the system's capacity to enter the pores and to migrate far enough to reach the defective layers, possibly passing through less porous layers without filtering out the solid component. It is obvious that just a few products can be considered appropriate to fulfil these requirements and that it may become necessary to apply a combination of products, if possible members of one chemical family, one after the other. Finally, it should be stressed here that the mode of application is of utmost importance.

Evaluating the effect of consolidation is again a difficult task whose major aim is to make sure that the weathering profile has been levelled out to some extent and, especially, no over-strengthening of any of the layers has been produced.

The most frequently used stone consolidants nowadays comprise ethyl silicates, inorganic micro- or nanosuspensions based on lime or silicate and synthetic resins in the form of solutions or emulsions. None of these products can be considered appropriate for all lithotypes and the full range of weathering defects. Considerable research efforts are needed to optimise the consolidants on the market, define their applicability for specific cases and develop new systems for cases of application where the existing treatments do not work properly.

2.1.4 Defects and Failures on Historical Mortar*

Several phenomena described earlier and concerning especially sedimentary stones are typical also for mortars, or generally artificial quasi-brittle composites, if they are present in a historical building in bulk form, that is, mostly inside masonries, including joints.

Surface layers made with mortars have specific features, and here the stone degradation glossary may not be sufficiently descriptive. Typically we understand the term 'surface layer' as a layer, or a composed system of layers, firmly attached to a substrate which usually possesses different material quality and characteristics. From the nature of monuments it follows that the composed layers may also include parts of original material which was on a surface transformed by means of physical or chemical changes.

So far no appropriate terminology and definitions for surface layers and their defects and failures have been introduced or codified and used. Therefore, this section outlines a brief systematic classification of defects and failures of plasters, claddings and paintings on plasters (including paints). A very short review is also included, describing investigation methods suitable for assessing the causes of their origin.

As has already been mentioned, ICOMOS has issued a glossary of defects [2], damage and failures of stone surfaces. However, this is only partially applicable for the description of defects on surface mortar layers. Nevertheless, the glossary is still very useful as a tool for a systematic description and classification of defects, damage and failures of surface layers. The detailed division of the stone glossary section, however, is less applicable as it is over-complicated and thus not practical for everyday inspection of historical objects. Moreover, for the purpose of this book, a further step consisting of the identification of possible defect initiation is considered more important than a very detailed complex phenomenological description.

---

*This section is written by Miloš Drdácký.

Taking into account the fact that the surface layer defects and failures are closely connected to their structural arrangement, the classification system distinguishes failure of:

- Thin surface layers
- Single-layer plasters
- Multilayer plasters
- Lining layers and mosaics
- Plasters (gypsum) with plastic decorative elements

The defects and failures can be sorted into seven basic groups:

- Cracks
- Deformations
- Detachment of layers from the substrate
- Damage due to material degradation
- Damage due to loss of material
- Loss or change of colour and surface deposits
- Biological layers

The last two groups can be defined according to the stone glossary because of their identical manifestation and no special approach to their assessment or remedial intervention is necessary, except in the case of the problems associated with wall paintings.

It might be possible to develop a specific group for surface failures induced by substrate failures, especially by failures and defects of supporting or bearing structures. However, in our system such phenomena are considered among the surface defects, too.

In addition, the combination of various types of defects is also very common, for example, cracks and detachment, which may then generate a series of variants in the defect morphology. There can be particular defects characterised by loss of cohesion or blistering in closed or open forms, smooth shapes or folded.

Even though it is not the aim of this chapter to review methods of investigation, documentation and study of surface layer failures, we can examine some basic data. The location and extent of sub-surface defects can be investigated by means of non-destructive or considerately destructive methods. Passive and active thermography can be used [3, 4] as well as naturally (environmentally) excited active thermography [5]. The defects location can be more

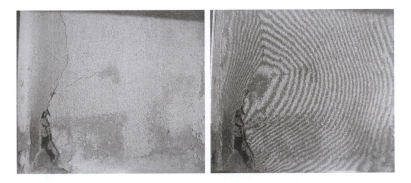

Figure 2.1 Moiré topography of a deformed detached lime plaster where the detachment is combined with the disintegration of the first-coat plaster layer and loss of material in the fold hinge. The form has developed for 35 years. The contour lines correspond to 1 mm distance. Photos: M. Drdácký.

accurately measured by the use of differential surface vibrations of detached plaster areas. This is possible due to Doppler effect when the surface is excited by periodical acoustic pressure [6], by piezo-shakers [7] or electrodynamic shakers [8]. Acoustic tracing, in its direct or semi-automatic version, has also shown promise [9, 10]. A useful review of methods and tools suitable for exterior surface inspection is given in Ref. [11]. Other methods available for the detailed measurement of defects and failures include the shadow moiré method for surface topography (Fig. 2.1), laser profile metre, microdrilling for assessment of depth and defects and boroscopy for investigation of larger and deeper gaps.

Defects and failures of modern plasters, including the determination of their probable causes and recommendations for remedial measures, are discussed in the Guide Mineral Products Association 2012 [12].

The likely causes of failures are considered further in the detailed description of defects and failures, especially in relation to the material of the surface layer. There is more information in the professional literature expanding on the failures of plasters or other surface layers. They are mostly focused on modern construction systems, frequently with gypsum boards, which will in the future require restoration and conservation.

2.1.4.1 Surface layers

Thin surface layer: The term 'thin surface layer' is defined as a continuous layer of painting or paint/textile wallpaper which is connected to the substrate by natural cohesion due to chemical or physical forces or through another thin layer, for example, of glue, ensuring adhesion. Thin surface layers may be created as one layer of one material or multilayer system of different materials. The usual thickness does not exceed 2 mm.

 Single-layer plaster: 'Single-layer plaster' is defined as a surface homogeneous layer of matured mortar which was created by throwing mortar mixture on a substrate at one sequence during plastering operations. The usual thickness does not exceed 20 mm.

 Multilayer plaster: 'Multilayer plaster' is defined as a surface quasi-homogeneous layer of matured mortar which originated either naturally by creation of a surface crust on a single-layer or multilayer plaster or by placing—usually different—mortar mixtures on a substrate at two or more sequences during plastering operations. The usual thickness does not exceed 25 mm; however, there are examples that exist of thicknesses up to about 100 mm.

 Lining layers and mosaics: 'Lining layers and mosaics' are defined as surface heterogeneous layers consisting of ceramic, china, glass, stone or natural elements such as tiles, tesserae, pebbles or shells embedded in a multilayer mortar substrate. The usual thickness does not exceed 40 mm.

 Plasters (gypsum) with plastic decorative elements: These are defined as a surface homogeneous layer of gypsum plaster which was created by throwing gypsum mortar mixture onto a substrate or as a final layer on a first coat of the plaster. It frequently creates a substrate for attaching decorative plastic elements which can be fastened (glued) or created in situ. The usual thickness does not exceed 20 mm.

2.1.4.2 Defects and failures: cracks

A crack is defined as an interruption of integrity or continuity of a surface layer by a linear gap. Cracks can be singular or multiple which originate in one point (star cracks) or create families of more or less parallel cracks, or they may create regular or irregular nets of cracks, for example a so-called crackele. Cracks can be fully developed—fissures or clefts (i.e., they penetrate through the full thickness of the layer and separate the integral undamaged parts)—or superficial, that is, without complete separation (they start on the surface, gradually weaken and ultimately vanish inside the layer). According to further characteristics or causes of their origin we can distinguish:

Hair cracking: This is defined as an interruption of the integrity or continuity of a surface layer by a linear gap of a thickness less than 0.1 mm.

Fine cracking: This is defined as an interruption of the integrity or continuity of a surface layer by a linear gap of a thickness between 0.1 mm and 1 mm. The cracks may originate due to various reasons, and there is no need for their specific treatment in buildings because they do not endanger their stability; however, they usually signal structural cracks in the substrate.

Technological cracks: These typically originate in places of so-called working joints or as a result of volumetric change and tensions in chemical processes of creation or metamorphoses of surface layers—they include shrinkage cracks. After the setting they are usually stable and can be repaired with unreinforced or reinforced local bonding. This defect is mostly caused by too 'fat' mixtures or very fine aggregate in mortar layers. In paints they are a result of membrane stresses due to shrinkage after chemical reactions and volatile compound migration.

Dilation cracks: These are typical for surface layers that include materials in which temperature and/or moisture dilation generate significant volumetric changes or different behaviour in relation to the substrate. In contrast to technological cracks, dilation cracks are not stable. The cracks are old and usually dirty which underlines their appearance. As they represent a natural system response it is not recommended to firmly 'stitch' them or fill them with stiff material.

Structural cracks: These are usually larger cracks which always penetrate across the full thickness of a surface layer and continue

Figure 2.2 Fine cracks as a result of dilation movement of the stones in a wall. In such a case the crack development cannot be prevented, and this defect may only be treated by using a mortar layer reinforced with a mesh or by regular repainting to diminish its appearance. Photo: M. Drdácký.

Figure 2.3 Craquelé is a typical pattern of shrinkage cracks. On the right side combined with detachment, deformations and loss of material. Photos: M. Drdácký.

into the substrate—where the crack continues into the substrate, it need not exactly coincide with where the surface layer was detached from the substrate. Such cracks are sometimes called static, even though they may originate due to different types of mechanical action, including dynamic or fatigue phenomena. They may be stable or unstable dependent on the bearing structure behaviour.

Figure 2.4 Combination of dilation and structural cracks due to the different expansion progress of the wooden ceiling and the plaster layer and also sag of the wooden floor structure. Photo: M. Drdácký.

Discovering the origin of the cracks usually requires long-term monitoring of the crack movement.

2.1.4.3 Deformations

Local swelling or shrinking: There are local surface layer defects in form of bulges, burles or craters that have originated within the layer without separation from the substrate.

Figure 2.5 Structural crack in stone masonry manifests in a plaster layer. Photo: J. Adámek.

Figure 2.6 Craters originated within the layer. Photo: M. Drdácký.

Buckling: This is deformation of a larger integral plate of the surface layer in a shape of smooth or folded (saddle roof) buckle or bulge, always combined with detachment, due to compression forces within the layer (e.g., Figs. 2.1 and 2.7).

Figure 2.7 Buckling patterns, one with the partial loss of material (upper right). Photo: M. Drdácký.

Bowing: This is deformation of a larger integral plate of a surface layer in a shape of smooth buckle or bulge, usually combined with detachment, due to a loss of stiffness or loss of strength.

2.1.4.4 Detachment of layers from the substrate typically combined with deformation

Blister: This denotes a raised portion of a surface layer in a shape of smooth or folded (saddle roof) form—which may be closed or with open peaks—and accompanied with a local loss of cohesion with the substrate or integrity of the layer under the blister (e.g., Fig. 2.8).

Blossoming: This creates a local deformation of a crack's separated surface layer in a flower leaf-like form with wound leaves accompanied with a loss of cohesion with the substrate along the edge and later the whole area of the detached leaf. Conservation intervention depends on the cause of damage—in thin layers typically a residual stress in the surface layer, and in the mortar layers blossoming tends to be caused by expansion of the detached interface layer.

Cupping: This creates a local deformation of a crack's separated surface layer in a form of a cup, accompanied with a loss of cohesion with the substrate along the edge. It has the same causes as blossoming.

Figure 2.8 Closed and open blisters in the surface layer locally detached from the first coat. Photos: M. Drdácký.

Figure 2.9 Deformation of cracked and detached layers in convex bending of the released leaves is classified as blossoming. Photos: M. Drdácký.

Lamination (exfoliation): This represents a defect in the splitting of individual layers in a multilayer system (e.g., Fig. 2.10) or the splitting of a layer into separated thinner layers—it is rather typical for multiple-layer systems produced with technological breaks or as a result of frost action.

2.1.4.5 Detachment, or loss of adhesion of a layer to the substrate

Detachment is an interruption of continuity and integrity near an interface between a substrate and a surface layer (Fig. 2.11). The surface layer's response to knocking has a typical 'dull' sound; it can

Figure 2.10 Combined damage of lamination of different layers of paints and fine coats with loss of material (left) and frost-induced lamination (right). Photos: M. Drdácký.

be seen in the form of a bursting of an integral thin floe in very hard and strong paints or plasters (Fig. 2.12).

Peeling off: This is a special loss of cohesion with a substrate similar to the blossoming but to a larger extent, typically seen in thin layers on ceilings and vaults.

Folding of an integral layer: This denotes a pushing down of a detached integral layer into a system of folds.

Figure 2.11 Detachment of large floe of plaster due to interface layer degradation and disintegration. Photos: M. Drdácký.

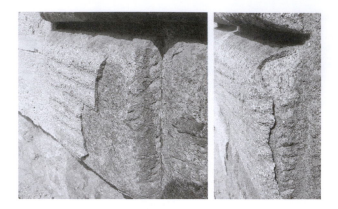

Figure 2.12   Detachment and spalling of stone plastic repair layer. Photos: M. Drdácký.

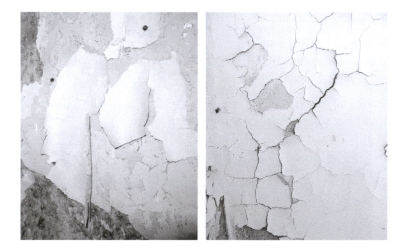

Figure 2.13   Peeling off thin layers. Photos: M. Drdácký.

Flaking: This is a combination of a dense crack net and a loss of cohesion with a substrate or a fine lamination.

2.1.4.6   Damage due to material degradation

Powdering: This means a loss of cohesion inside the material, with degradation and disintegration of a composite matter into very fine grains—it is typical for thin surface layers.

Figure 2.14 Loss of cohesion (sanding) of the first coat plaster under a hardened surface crust layer (left) and disintegration of plaster with flaking (right). Photos: M. Drdácký.

Figure 2.15 Loss of material by scratching off on a wall which is deliberately used as a support for bicycles (left) and craters made by impact of bullets (right). Photos: M. Drdácký.

Loss of cohesion, disintegration and crumbling: This means a loss of cohesion inside the material, with degradation and disintegration of a composite matter into aggregate grains—it is typical for mortar surface layers. In multilayer systems the disintegration frequently occurs in the first coat of weaker mortar below a harder and more

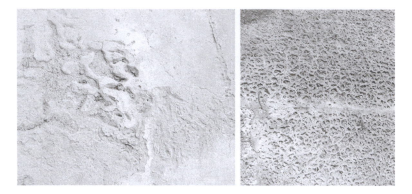

Figure 2.16 Selective erosion (left) and alveolisation (right). Photos: M. Drdácký.

Figure 2.17 Selective degradation and erosion. Photo: M. Drdácký.

resistant surface layer as a result of selective degradation along the depth profile.

2.1.4.7 Damage due to loss of material

Loss of material, erosion and wear: This is caused by various reasons, including washing away, rubbing off or scratching in thin surface layers (e.g., Fig. 2.15). In addition selective erosion (e.g., Fig. 2.16), pitting, lacunae can be caused due to natural deterioration processes or intentional mechanical damage in mortar layers.

2.2 Physical Properties and Their Characterisation*

The characterisation of physical properties of existing substrates represents one of the most important survey activities preceding any intervention aimed at consolidation, strengthening, injection and even non-structural work on historical objects. It may take advantage of all known testing methods and practical techniques that are routinely used in this field of science. Characterisation is also supported by various standards or codes and recommendations. However, in the case of historical objects the standards frequently cannot be fully applied due to the impossibility of ensuring the bulk of material necessary for testing specimens in a prescribed size, dimension and quantity. Therefore, non-standard, direct test methods or indirect non-destructive (NDT) or considerately destructive (also medium destructive [MDT] or semi-destructive [SDT]) methods have been developed and are frequently used.

There are several reasons for recording the physical properties of historic substrates. Firstly they provide the restorer with the data necessary for the selection of appropriate materials and technologies compatible with the historical substrate from a physical perspective. In addition they help to provide evidence to inform the initial decision concerning the feasibility of any intended intervention, for example, the degree of penetrability of the selected agent into the treated material. In situ tests are also important for quality control and assessment of the treatment efficiency. Finally, the physical testing creates a basis for long term monitoring of historical objects.

In the field of conservation the following properties are typically investigated:

- Material density and specific weight
- Specific porosity (microstructure)
- Mechanical characteristics
- Wave propagation
- Water transport and retention

---

*This section is written by Miloš Drdácký.

- Thermal and hygric/hydric properties
- Material colour

This chapter is only concerned with properties for which non-standard techniques or in situ methods have been developed.

2.2.1 Testing Material and Test Specimens

Material characterisation is always referred to actual objects and the restoration practice, these historical items are often associated with cultural heritage and as such usually have some level of protection. Therefore, the availability of material samples for testing is usually very limited. Exceptions to this may be found in cases of failures due to natural disasters, partial demolitions, approved modifications concerned with replacing some parts of the monument and also the recovery of spare material for future repair (e.g., roof tiles). In such cases the historical material may be available in the bulk required for manufacturing test specimens in the quantity necessary for the general standards used in the technical testing of materials. Nevertheless, even in these exceptional cases there is usually a shortage in the amount of material needed to generate the number of specimens required for statistical evaluation. An example of this concerns roof tiles, where it is usually possible to obtain a few degraded specimens which are to be removed from the roof and prepared for repair. However, these will still not generate the amount of specimen material required by the majority of standard testing methods. Fortunately, in specific cases, it is possible to take advantage of the standards that exist for assessing the condition of historical materials and objects, which justifies the application of non-standard approaches.

The supply of material from protected cultural heritage objects for testing must comply with the relevant international standards concerning interventions into historic tissue. European technical standardisation committee CEN/TC 346 'Conservation of Cultural Heritage' has prepared at least two standards which are applicable. The EN 16096 standard (August 2012) on 'Conservation of cultural property – Condition survey and report of built cultural heritage' requires that assessment of the condition of historic objects should be based on visual inspection combined with appropriate simple measurements. The following EN 16085 (August 2012) 'Conservation of cultural property – Methodology for sampling from materials of cultural property – General rules' defines basic approaches to the extraction of material from historic structures, buildings or other objects. No specific rules are suggested for the number of samples taken and all interventions must be individually considered by the assessor in the sampling plan, taking into account the condition of the structure and the objectives of the sampling procedure. Minimum intervention is a general rule and correlation with other NDT methods is recommended. In other words, the SDT, that is, direct techniques, should preferably be used for calibration or to prove estimates are achievable by indirect NDT methods.

Non-standard methods then, provide us with the tools which enable testing of physical properties on small samples, non-standard specimens or on real objects. Therefore, such methods are discussed in further detail here.

2.2.2 Mechanical Characteristics

Basic mechanical characteristics which are important for the planning and implementation of restoration interventions, involve the strength characterisation of the historical substrate, the modulus of elasticity and the Poisson ratio. Reliable data can be achieved by means of destructive as well as non-destructive testing on laboratory specimens or in situ on the actual objects. It should be noted that from the list of mechanical characteristics outlined above, not all are important in all situations. Their importance is related to the real loading and environmental conditions of the cultural heritage object being assessed. The strength of the substrate is usually investigated for an assessment on the degree of degradation and the efficiency of consolidation treatments. The modulus of elasticity and the Poisson ratio are important for mathematical modelling of materials. The modulus of elasticity also plays an important role in maintaining the compatibility between the original and the repair material or material strengthened after consolidation. However, this is only valid in cases where deformation differences in the composed material or structure may initiate strains and stresses threatening its integrity. Abrasion resistance, toughness and fatigue resistance, drilling resistance, hardness, and sub-surface cohesion are of concern in specific cases, and thus can be used in any assessment of the substrate state before and after conservation.

2.2.2.1 Non-standard destructive testing of the strength characteristics of historical substrates

2.2.2.1.1 *Compression test*

Compressive testing of mortars in historic masonry is problematic. It is clear that standard mortar tests yield more or less meaningless results, which are practically useless for assessing the current safety of, or the potential threat of damage to a historic structure. Mortar in historic structures is not subjected to compression in a way similar to the standard test conditions and, therefore, alternative non-standard approaches have been developed [13].

The methodology for tests on small specimens is influenced by several factors. Firstly, by the fact that the real size of a mortar sample taken from a historic structure – from the masonry – is usually less than 20 mm in thickness. Manufacturing the specimen for a compression test (cutting a cube) has a significant influence on the properties of the sample, as it basically disturbs the surface strata and reduces its strength. In addition, manufacturing a small cube is laborious and time-consuming. In most cases, the compressed surfaces need to be supplemented with a levelling layer. For this reason, specimens in the shape of irregular mortar 'cakes' from the masonry joint have also been studied for compression tests in recent times, for example [14].

However, it has been known since the 19th century, that the size of the test specimen has a significant influence on the measured strength. Numerous methods for adjusting the attained characteristics of concrete and cement mortars to the standard values have been proposed. It has been concluded that the size effect correction adjustments for concrete depend mainly on the length of the specimen base edge, on the slenderness or the height-to-base edge length ratio and on the quality (compression strength) of the mortar. The maximum grain size of the sand is not as important as other factors, for example, the testing arrangement eccentricity and/or testing frame end plates, and their friction characteristics. The strength attained on non-standard samples is higher than the strength measured on standard specimens, if:

- The height (or thickness)-to-base-length ratio decreases
- The base length decreases
- The standard compression strength decreases

However, little work has been done on pure lime mortars and on specimens less than 2 cm in thickness. A review of results of tests on cement mortars, based on a literature survey supplemented by a series of tests on a cement mortar [15] was published by Drdácký (2011) [13], who developed correction coefficients applicable for assessing the equivalent standard compression strength from tests on non-standard specimens.

For lime mortars with very low standard strength, that is, close to 0.365 MPa, which corresponds to degraded historic mortars, the experimentally measured strength of specimens with low slenderness can be corrected with the use of Eq. 2.1

$$f_c = f_e/(h/a)^{-1.9114} \qquad\qquad (2.1)$$

where f_c denotes the computed equivalent standard compression strength, f_e the experimentally obtained compression strength, h/a the slenderness of the specimen. The formula is valid for specimens with length a of the base side of about 40 mm and with low strength (the lime : sand mixture was only 1 : 9 vol%). Equation 2.1 is compared to experimental results in Fig. 2.18. It is worth noting that in the tests of lean lime mortar no specific friction conditions between the compression plate and the loaded specimen surfaces were employed. The empirical form is dependent on the slenderness ratio of the tested specimen, its base length, its strength, the arrangement of the friction between the loading plates and the specimen, and on the type of test – punching a 'cake' like plate or compressing a specimen of the same shape and dimensions as the loading plates. Form (2.1) is therefore recommended for an approximate assessment of the standard equivalent lime mortar

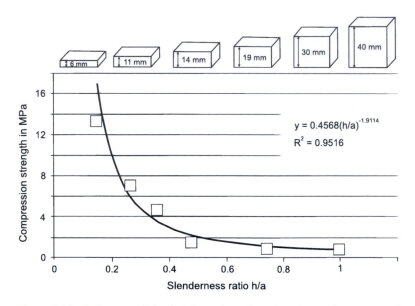

Figure 2.18 Influence of the height-to-base-length ratio on the measured compression strength.

qualities (compressive strength) applicable in standard redesign processes. For the lime mortars gauged with hydraulic additives the exponent reaches lower values, for example, -1.065 for lime-metakaoline-cement mortars.

Small-size non-standard specimens provide a limited opportunity to measure the deformation characteristics of mortars. Despite promising results from the successful application of optical methods (DIC) to determine the modulus of elasticity from compression tests on small-size mortar specimens [16], the most satisfactory results are still obtained from bending tests. It is therefore desirable to develop a method for the flexural testing of real mortar, together with any tension strength measurement.

2.2.2.1.2 *Bending Tests*
In bending tests, the specimens must be cut, which involves the same problem of surface disturbance. For joint mortars, at least two faces of the sample do not have to be used and, consequently their properties remain intact. These faces are always the contact

surfaces of the mortar and masonry elements and, consequently, these surfaces are influenced differently from the basic material due to the different setting and hardening conditions on the mortar–stone/brick interface. Nevertheless, the sample size is still too small for flexural tests. Therefore we have devised a sample extension with another material in the form of so-called 'prostheses', where the tested material is placed at the centre of the test specimen. The method is described in detail in Section 2.3 and illustrated in Fig. 2.35. According to the results of tests that compare the flexural strength of pure mortar beams with the strength ascertained on identical prosthesised material, the influence of prosthesisation is negligible. Specimens broken during flexural tests were used for testing the influence of prosthesisation on the results of flexural tests on mortars. Specimens with dimensions of approximately $2 \times 2 \times 3 \, \text{cm}^3$ and $3 \times 3 \times 4 \, \text{cm}^3$ were extended to the length required for flexural tests by gluing wooden prostheses to both ends which were then loaded in a four-point flexure. The values for the ratio of the strength of the prosthesised sample to that of 'standard', that is, all-mortar specimens, were recorded between 0.98 and 1.02 if the specimen broke in its central part (undisturbed mortar).

It was also necessary at this point to study the influence of specimen size on measured strength. The methodology was tested and calibrated using standard tests of mortar of various qualities and carbonisation values. The effect of size on bending strength was also evaluated from these tests. The results are summarised in Fig. 2.19, from which it follows that the size effect corresponds to the effect known from bending or tensile splitting tests on concrete or stone samples. Sand composed of three different size grain types (fine, medium and coarse) and three different cross-sectional dimensions (2, 3 and 4 cm long edges) were used. Seemingly, the grain size does not affect the small size reduction rate, although it slightly influences the strength of the mortars. The small size reduction tendency can be assessed by an approximate Eq. 2.2:

$$f_b = f_e/(1 + C_{ssb} - C_{ssb}h/h_s) \qquad (2.2)$$

where f_b denotes the computed standard flexural strength, f_e the experimentally-attained flexural strength, h/h_s cross-sectional height of the specimen to the standard $h_s = 40 \, \text{mm}$ ratio, and the

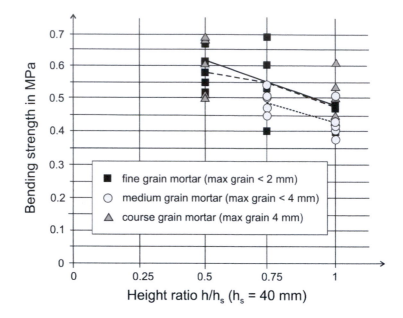

Figure 2.19   The effect of beam cross-sectional size on bending strength.

small size bending correction coefficient C_{ssb} varies on an average around 0.47 for lime mortar specimens between 20–40 mm in height (the measured C_{ssb} varied from 0.36 to 0.58 and apparently depends on the technological parameters, for example, compacting and curing treatment). The modulus of elasticity can be estimated from the measured deflection of the beam during the course of loading.

2.2.2.1.3 *Shear tests*

The non-standard compression and bending tests are accompanied by shear tests derived from the methodology for the shear testing of soils. For tests of this type, a specimen of any convex shape is embedded into a block of stiff material, for example, epoxy resin divided into halves (Fig. 2.20). The dividing plane is provided with a separation and a sliding layer. Then the block is tested in a simple shear box and the measured normal and shear stress is evaluated using the standard soil mechanics methodology. The broken parts of the mortar beams from the bending tests can be advantageously reused in the shear tests.

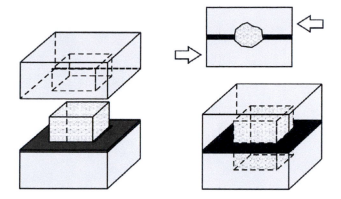

Figure 2.20   The principle of shear tests on mortar specimens.

2.2.2.1.4 *Abrasion tests*

Mortars for pavement and floor surfaces are tested for resistance against abrasion. Again it might not be possible to extract standard test specimens in the shape of a cube $71 \times 71 \times 25$ mm^3 from real structures, therefore, usually thinner plates are glued to supporting blocks for fixing in the abrasion machine. For the Böhme test a specimen is pushed against a moving grinding wheel with constant force. A loss of material after a defined number of rounds is then evaluated [17].

2.2.2.2 Non-destructive testing of mechanical characteristics

The purely non-destructive methods are based on measurements of the response of an object's material to mechanical excitation or loading. From the rather large family of these techniques only the most useful and applicable are presented below.

Ultrasonic investigations proved to correlate very well with the modulus of elasticity and the strength of material. Basically, the velocity of transmission of ultrasonic waves is measured which correlates with porosity of the investigated material (Fig. 2.21). This NDT method is standardised (for example, the European Standard EN 14579:2004), and very useful, but any evaluation of measurements involving materials and objects of a cultural heritage form, would be affected by so many different circumstances that it is

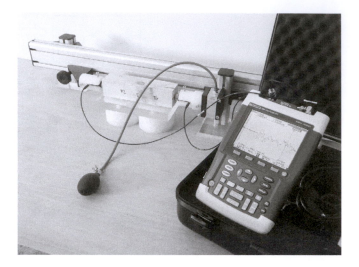

Figure 2.21 Testing of a beam sample with ultrasonic waves transmitting transducer (left) and receiving transducer (right).

beyond the scope of this book. More detailed information about the method is explained in Ref. [17].

A group of MDT or SDT techniques is usually included among the non-destructive methods. They have been developed in many cases specifically for use in the areas of restoration and conservation. They supply the restorer with mechanical characteristics which correlate with classical mechanical material qualities.

The resistance of historical substrates to the penetration of a drill has been used for the assessment of its strength and cohesion for a considerable time. Several different methodologies exist to perform this task and there are also a range of devices for resistance measurement available on the market. The simplest method is based on the measurement of the depth of penetration of a standard drill after a given number of revolutions. The method has been calibrated for bricks and sandstone and it provides good data as long as the drill bit is sharp. This can be tested using a calibration sandstone probe supplied with the device. The drill operator needs to be practised in applying a constant pushing force on the drill penetrating the material. More sophisticated devices measure electrically the resistance of the material against drilling

Figure 2.22   Resistance drill testing of a medieval plaster.

at a constant axial pressure on the drill bit. The older variants, for example, [TERSIS – Geotron Elektronik, Germany] generate the pressure by using weights (Fig. 2.22) which limit its practical operation. The most advanced development takes advantage of a pneumatic pressure which accommodates application of the method in any position and direction of drilling without the necessity of fixing the device to a tripod. Another device measures the drilling resistance and the force of pushing the drill into the material [DRMS – Sint Technology, Italy]. This device is both light and user friendly [18].

Drilling resistance methods give satisfactory results mainly on homogeneous and isotropic materials, for example, marble. On stone and mortar composed of hard grains in a rather soft matrix, the results are not so reliable if the length of drilling is too short. Del Monte and A. Vignoli (2008) [19] recommend at least 30 mm depth for mortars. Therefore, for the case of shallow surface degradation or consolidation of mortar or sandstone the method does not provide representative data. In some cases it provides no useful data, even if it is used for relative measurements – before and after consolidation.

The shallow subsurface cohesion can be reliably measured by using the peeling test ('Scotch tape' test). This procedure is reliant on being able to measure the released material directly on site, or to transport the samples to a laboratory. In either case, on site or at the laboratory, the first step is the same. A clean, dry section with no gross defects and imperfections which is representative of the material to be tested is selected. Scotch tape is then placed onto the selected surface. If the sample is transported to a laboratory for further processing, the use of double tape samplers is recommended, which have to be prepared in the laboratory before the in situ test is performed. These samplers can be rather small in size, approximately 20 mm by 40 mm. A plastic strip with an adhesive layer on one side only can also be used if it is long enough to be folded for transportation with the attached particles.

During the measurements, the double tape sampler has to be taken out of the sampling bag, and then the cover sheet of the adhesive layer must be removed and kept for readjusting after the peeling has been done. Next the tape or plastic strip is stuck to the surface, see Fig. 2.23. The tape is now smoothed into place using a finger, and after approximately 60 seconds the tape is removed by taking the free end and pulling it off steadily at a rate of about 10 mm/s and at an angle of 90°. The protective sheet is placed over the adhesive layer with the released material attached and inserted back into the transportation sampling bag, finally sealing by closing the zip. When a single sided plastic strip is used, there is the problem of measuring the amounts of attached material, which can be done in situ by a portable balance. As the balance is such a sensitive device its operation in the field may be too impractical

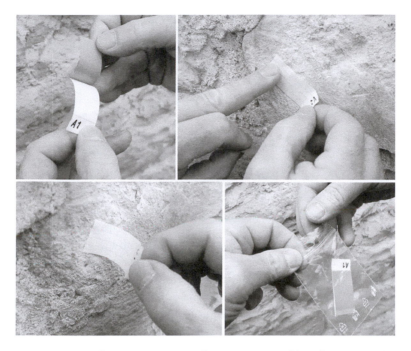

Figure 2.23 Peeling test sequence of operations: Double side tape sampler with the cover sheet of the adhesive layer removed (upper left), the tape or plastic strip stuck and gently smoothed to the surface (upper right), tape sampler removal from the tested place (lower left), double side tape inserted back into the transportation sampling bag (lower right).

or even impossible, for example when working on scaffolding. Therefore, it is suggested that the tape is carefully folded after measurement by simply bending it along the centre and placing it in a transportation box specifically designed for this purpose. The sampler can then be transferred to the laboratory (Fig. 2.24) where the weight can be recorded. When using a double tape sampler, it is advantageous to weigh the bags and samplers together before attempting measurements in the field, and then repeat this when returning to the lab with the samplers carrying the released material. Plastic strips are usually not weighed before application, because the differences in the thickness and weight of the plastic strip are more or less negligible in comparison with the mass of the released sand or stone material. The folded strips with the attached

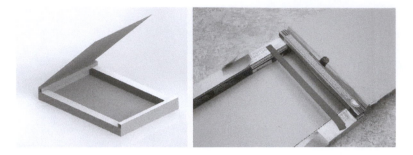

Figure 2.24 A simple paper box for transporting the tapes with the attached material to laboratory.

material between the two strips are only cut to a given length in the laboratory and weighed for a comparative evaluation. This technique simplifies and accelerates the measurement procedure, and makes the method more versatile in contrast to previously published recommendations. Instead of plastic strips, commercially available paper labels with adhesives may also be used, provided there is no possibility that they will become wet [20, 21].

On site, the test should be repeated 10 times, ideally at the same position. The procedure should then be repeated in other suitable places on the surface being investigated. This repetition avoids the biasing of results because the loosened particles on a fresh tested surface do not represent the 'near surface cohesion' characteristics of the tested material. If we repeat the peeling in the same place we observe a decrease in the amount of material released and thus an apparent, but false, consolidation effect. Obviously, the same problem exists when consolidation treatment effects are checked on degraded historic materials or on deteriorated materials of any other age. After sticking and peeling has been repeated several times, the amount of detached material starts to become almost constant and characterises the cohesion of the material (Fig. 2.25). The illustrative figure clearly shows that the method is sufficiently sensitive to lime mortar quality variations caused by different mixture composition.

To evaluate the weight measurements, a non-linear approxima- tion model is adopted. This is necessary if the data converges to some positive, that is, non-zero, value. The approximation function

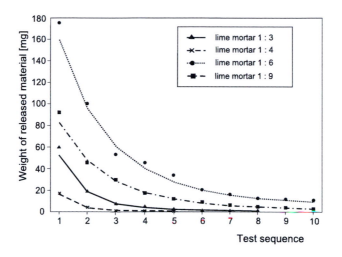

Figure 2.25 Weights of the released material particles at individual test repetitions on lime mortars of various lime to sand ratio (1 : 3, 1 : 4, 1 : 6 and 1 : 9).

in the shape presented in Fig. 2.26 relates the amount of the released material ($m(n)$) in a peeling sequence to the number of the peeling sequence (n). The determined approximation asymptote is considered to be a surface cohesion characteristic. This value A [g] correlates quite well with mechanical characteristics; an example of measurements on various types of sandstone is shown in Fig. 2.27. The method is also very useful for comparative measurements and quality assessment of consolidation effects.

Another measurable parameter may be the peeling force. A device has been specifically developed by SINT (Regione Toscana) for resistance against peeling. This is obviously less practical, but it also corresponds quite well with the material characteristics.

2.2.3 Other Physical Characteristics

In the field of conservation and restoration, several other physical characteristics appear to be more important than mechanical characteristics. Firstly the sensitivity of a material to temperature and moisture change will affect any volumetric change of a material. Further, the porosity plays a decisive role in many aspects and

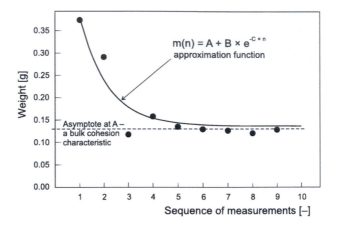

Figure 2.26 Non-linear approximation model for a sequence of measurements.

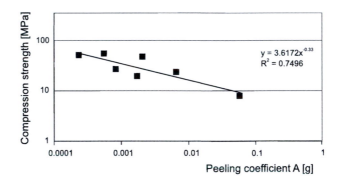

Figure 2.27 Correlation of a cohesion characteristic A with compression strength of various types of sandstone.

not only affects the treatability of the substrate but also its vapour and water permeability (wetting and drying cycle) and frost resistance. The material's colour and appearance are also usually recorded among the physical characteristics. Some of the applied testing methods for such a purpose are described in Section 2.3, for example, dilation or permeability characteristics. This part is focused on water uptake measurements which are indispensable in the conservation field because they provide a restorer with data important for assessment of the state of deterioration, the assessment of stone or mortar durability and treatability.

Water transport correlates with pore sizes and also with the mechanical characteristics of porous materials. In situ measurements of water uptake parameters take advantage of capillary water absorption which is measured in a standard way by means of so-called Karsten tube, which is recommended by RILEM. A Karsten tube is a scaled pipe-like glass tube, which is fixed with removable putty to the surface to be investigated. The tube is then filled with water (or a conservation agent) and the transport of water is followed on the scale and recorded together with time data. This method however, is not particularly user-friendly, and its application leads to many difficulties, namely problems with fixing a heavy glass tube to a vertical surface. In addition, it cannot be used for measurements on inclined or highly curved surfaces and ceilings, and the application requires two operators: one to observe the water movement in the scaled glass tube and operate the stop-watch, and another who records the readings. There can also be problems with sealing the contact ring area, and the sealing putty can soil the surface. Therefore, in the STONECORE project other techniques were studied and developed.

An innovative digitised microtube system for water uptake measurements has been developed. This technique is based on the need to: record the data electronically, record the water sorption from the start of the investigation, to make continuous measurements of water infusion into the surface, make measurements on arbitrarily inclined surfaces, and measure 'point' characteristics. No fixing to the surface and no special sealing are required. The mechanical part of this pistol-like device (Fig. 2.28) consists of a scaled capillary tube or any other scaled glass tube placed in a tube holder, which is adjustable in both horizontal or vertical planes (swivel connected), a connecting plastic hose, an outlet hinged head with a three-point support, a switch for manual data recording, and a connection cable to the data storage unit. The electronics are housed in a sheet metal box which has an OFF/ON switch on the front panel. The upper panel has an illuminated LCD alpha-numerical display, and in the upper right corner there is a two-position switch (with positions marked 0 and 1). This switch opens and closes the measured data group. The

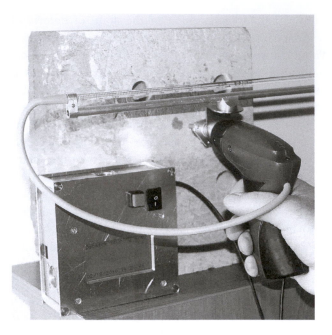

Figure 2.28   Digital microtube device for water uptake measurements.

red switch has the same function as the trigger on the pistol, and records an instantaneous time value. On the left side of the box there is a USB connector for transferring the data from the electronic unit to a PC or Notebook [22].

This unit records time intervals corresponding to the defined volume of water absorbed by the tested material. The water uptake velocity is observed on the scaled microtube, which is kept in a holder fixed to the pistol with a magnet. The holder also enables other tubes to be fixed, for example a Mirowski tube or even a Karsten tube (when rotated into vertical position) and the holding fixtures are replaced by fixtures that match the tube used. However, the horizontal capillary tube can be used for measurements on inclined surfaces for example, vaults or ceilings. The pistol trigger is modified in the form of a microswitch, which manually controls the recording of the instantaneous real time value. The microswitch is pushed when the water meniscus in the capillary tube crosses the scale line corresponding to a value of 0.01 mL in a typical capillary tube. The trigger is connected to the electronic unit through a cable with a connector. The device enables 40 measuring groups to be recorded, that is, measurements can be recorded at 40 locations. The records can then be stored on a computer so the device memory can be cleared and made ready for another series of measurements.

In the case of a microtube, the acquired data covers a very short period of water absorption, due to the limited amount of water in the tube. The records are therefore approximated by a linear function, which is used as the measurement representation instead of the usual water uptake coefficient. As an example, the Fig. 2.29 is a graphical representation of the measured data from in situ water uptake measurements on a differentially deteriorated 'opuka' (marl) stone wall – from a rather compact material (1), through slightly degraded rock (2) to seriously deteriorated and cracked material (3) and (4). Line (5) relates to a vertically cracked part of the relatively compact stone on which measurement No. 1 was made. It is apparent how cracks can influence the measurements. Further illustrative examples are included in the Chapter 7.

This microtube method provides both a practical and swift operation. The storage and manipulation of data is convenient, and measurements on arbitrarily inclined surfaces can be performed without surface soiling. It can be used for water uptake point measurements on very complex shapes with a small radius of curvature. It is more efficient than the Karsten tube method in terms of the number of personnel required and the time needed for surface water uptake characterisation. The device with a scaled glass tube is intended for measurements on surfaces with a medium rate of water uptake. For very high, as well as low, water absorption rates a version without glass microtube has been developed – here the contact sponge ('cigarette filter') is fed from a water container connected by a flexible hose and the uptake velocity is measured electronically by positioning a small float in the container.

The colour and appearance of cultural heritage surfaces is one of the most important physical qualities studied in their conservation. Typical point colour measurement methods have many disadvantages; they do not truly represent the appearance of the whole substrate surface and are therefore not suitable for long term monitoring. To overcome this issue, a full field appearance

Figure 2.29 A graphical representation of the data recorded from in situ water uptake measurements on a differentially deteriorated 'opuka' (marl) stone wall. (The photograph covers a length of 1 m.)

measurement technique has been developed [23–25]. In situ use requires a simple and cheap device that is at the same time portable and capable of detailed image resolution. A portable scanner is thus ideal for this technique. A scanning resolution of 600 dpi yields approximately 20 pixels per millimetre. Using a scanning area of $18 \times 13 \text{ cm}^2$, images of approximately 10 megapixels in size can be

Figure 2.30 A portable scanner used for examination of the colour and appearance of historic surfaces. Photo: J. Valach.

acquired. Taking into account the illumination stability, the portable scanner (Fig. 2.30) outperforms photography which is unsuitable for measurement purposes because of varying light condition affected by weather. Using a scanner, it is possible to measure surface visual properties and parameters of their statistical distribution reliably;

provided the image is calibrated by adjusting the highest and the lowest intensity levels against a black and white template frame included in every scanned image.

The measurement itself is very straightforward: on a sufficiently flat part of object's surface the scanner is positioned with the top cover removed. Once scanning is finished, the image is written on a SD memory card, or data can be sent directly via USB to a notebook. The use of a portable scanner and notebook for data processing (that can be carried out off-line), ensures complete portability and importantly the method does not rely on an external power source. Scanned images are then processed and analysed using a custom-written program running in MATLAB. The technique is also completely non-destructive.

There are several 'colour models' used to describe the colour properties of the substrate. Digital optical devices use a RGB (red, green, blue) colour model to record images, however, a HSL (hue, saturation, luminance) model is better suited to describe changes associated with surface degradation and thus transformation between the colour models has to be performed according to a set procedure [26]. Degradation processes at a CH monuments' surface are accompanied with a change in its appearance. Soiling of the surface manifests itself visually as a darkening of the surface and loss of its colourfulness. The visual changes described above can be also communicated in the more precise technical terms of brightness and saturation. Brightness is a measure of the amount of light reflected back from a surface. The more light that is reflected back, the higher the brightness value. The two limiting states – black and white are ascribed values 0 resp. 1. Saturation is a measure of colour purity – at one end of the scale are the pure rainbow colours and at the other end colours with additions of black, white or their mixture – grey. The fact that the judgement can be made independently of the illumination conditions shows that variation of brightness and saturation are properties that depict the visual state of the surface with the high representative validity. Pixels making up the image of the scanned surface, are then statistically processed – mean values and standard deviation for brightness and saturation are determined and used for further analysis of the surface's optical characteristics. The mean

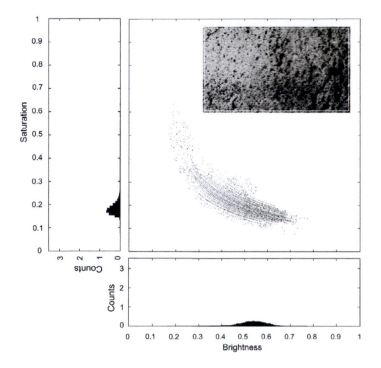

Figure 2.31 A typical saturation/brightness coordination plot for a soiled surface.

value represents centre of gravity of all the points' positions on the brightness or saturation axes. The method outlined analyses parameters of the statistical distribution of these properties to approach quantitatively what is meant by an 'uneven degraded surface'. The shift of the mean values of these two variables indicates an increase in surface degradation or can be used as a measure of any cleaning effectiveness. Properties of statistical distribution can be visualised by using a saturation-brightness coordinates plot, in which every pixel of the studied image is represented by position of the brightness-saturation coordinates. Conversion of the image's pixels into brightness-saturation coordinates creates a 'cloud of points', its dispersion and shape being a useful measure of surface degradation (Fig. 2.31). A dense cloud means uniform appearance, while dispersed points signify more variation at the surface.

2.3 Physical Modelling and Testing of Consolidation Effects*

2.3.1 Assessment of Consolidation Effects

The consolidation of stone and mortar represents one of the most complicated restoration technologies. This type of intervention will almost certainly influence the material qualities of a historical object. Conservation science strives constantly to produce new agents for stone and mortar conservation which are in heavy demand in today's marketplace. However, their usage in conservation projects requires the careful investigation of their impact on the historical material from a sustainability and conservation point of view.

Any assessment and evaluation of the impact of new agents on cultural heritage is a difficult task, where there is usually no real desire to study all the possible effects or to obtain absolute standard values because the relative data is mostly sufficient for decision making. Ideally, the effects of any remedial application should be investigated using real materials under real conditions to assess any immediate impact on the substrate, and to predict any potential future consequences. In reality, the investigations are mostly performed in laboratories with some rather limited complementary checks on the real objects in situ. In fact, appropriate testing facilities are usually not available for the exact determination of some characteristics in situ. Therefore, the most appropriate tests are usually set-up to study the more critical effects, which may compromise the compatibility of the intervention procedure with the historic material and its condition and operational environment. These tests typically involve the investigation of the mechanical and physical impact due to any consolidation intervention.

Tabasso and Simon (2006) [27] published a critical review of the test methods developed for the assessment of stone properties before and after any conservation treatment. This review provides useful information on both standard and non-standard techniques. The investigation of stone substrates and the evaluation of conservation intervention procedures, including recommendations

---

*This section is written by Miloš Drdácký.

for establishing criteria for compatibility of applications for specific projects have been studied comprehensively in Germany [27, 28]. Work in Germany has also been very innovative, with the development of many new methodologies. However, very few reliable methods have been published for the testing of consolidation effects on mortars. Many of the problems associated with the development of these methods are related to the diversity of materials and their deterioration. Tackling these issues requires a creative approach to the physical modelling and testing of consolidation effects and some examples are presented here, whilst later chapters have focused on application techniques and case studies.

As previously stated, the assessment program must be undertaken in the knowledge that original historic material from the chosen monuments will rarely be sanctioned for any form of destruction to allow the extraction of a large bulk of material. Therefore, the consolidation effects are typically studied on model materials or in exceptional circumstances on authentic materials which are to be replaced due to their state of deterioration. However, even these original materials may not be representative for any consolidation assessment due to the wide variation of their characteristics.

The testing of stone consolidation is usually carried out based on various standards recommended through investigations on artificially weathered stones, for example, Italian NORMAL 1985 [29]. Several techniques have been developed for artificial weathering, for example freeze-thaw cycles, heating and salt crystallisation cycles. However, for the purpose of comparative testing to establish the efficiency of different consolidants, a simplified approach based on the use of model materials is considered acceptable.

The testing of consolidation effects aims to obtain comparative data with minimal errors and thus appropriate homogeneous samples with only slight variability in important parameters are required. In the case of stone, a basic series of tests are performed on a sound material preferably extracted from historical sources. Stone can be readily cut for the manufacture of test specimens. However, this method is not suitable for the generation of mortar specimens. For investigations involving the consolidation of historical mortars, it is more feasible to use newly prepared specimens from fresh mixtures or dry compacted aggregates.

Another specific problem to be overcome in consolidation studies is the capacity of the test substrate to absorb the impregnating agent. Penetration depth is one of the most critical testing parameters, which influences the planning of the test significantly. Typical penetration values for consolidation agents applied onto stone vary from depths of several millimetres to several centimetres and are dependent on the size and connectivity of pores.

Any testing program combines both destructive and non-destructive tests and it is desirable to test individual parameters on very similar specimens. It is essential that this requirement is considered when initially planning a sequence of tests. In addition, the maturing time of both the substrate and the consolidating agent must be taken into account. Typically the procedure begins with the non-destructive tests and continues with destructive mechanical tests after which broken parts are used for further investigations. The sequence is outlined in the following sections.

2.3.2 Test Specimens for Material Testing

The practice of using thin circular plates for the testing of consolidation effects on stone has been common for decades [30]. This method is advantageous for studying effects on samples procured by core drilling of real structures. A typical radius of a circular support is 18 mm, with a a loading circle of 6.5 mm [31], or this is modified for samples of 80 mm in diameter [32]. For extensive laboratory testing, the cylinders produced by this method are not practical and testing consolidation effects on plates cut as cubes is more efficient (Fig. 2.32) [33].

The cube is treated on one surface by the application of a consolidation agent and after maturing, the penetration depth is checked by collecting ultrasonic measurements across the cube width. Subsequently the cube is cut into two parts – one part is used to prepare thin slices which are used as flexural test specimens. The other part is used to produce specimens for compression tests. Cutting thin plates also enables the study of stone samples

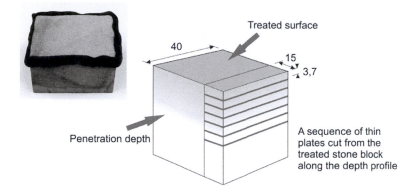

Figure 2.32 The cutting of thin plates from partially impregnated cubes.

taken from real structures with surface deterioration [34]. This is probably the most reliable and informative method for describing any variation in stone characteristics. Compressive tests on stone have not been used frequently for testing consolidation effects. This is mainly because this type of test has the associated problem of not achieving full and even penetration of the consolidation agent throughout the entire volume. Therefore any information gathered is of limited use.

The approach outlined above for assessing the consolidation effects on stone cannot be used on mortars as the reliable cutting of thin plates is almost impossible. For testing mortars, thin plates for tension tests are cast, and for compression tests two options are possible. Short thin-walled tubes that have full and even penetration can be used or an alternative is to utilise small blocks in which the agent distribution need not be complete but is even and regular (Fig. 2.33). The tubes can easily be treated with many cycles of impregnation when placed on a simple rotational shaft (Fig. 2.34). The casting of special specimens enables the design and production of mortars of various qualities imitating the real substrates. For example, a degraded lime mortar has been treated with a weak (lean) lime mixture prepared from lime binder and sand in ratio of $1:9$ [35]. This lean (weak) mortar increases the

Figure 2.33 Thin walled cylindrical (left with a formwork) and plate (right) specimens. Photos: M. Drdácký.

Figure 2.34 Treatment of cylindrical specimens on a rotational shaft. Photo: T. Drdácký.

sensitivity for observing any strengthening or consolidation effects. The attenuation of the tension plate apparent in Fig. 2.33 can be omitted, especially for short plates. Flexural tests are not needed on these tension and compression specimens.

Degraded mortars or even stone can be modelled using dry compacted crushed aggregate of a given grain distribution. The main advantage lays in the fact that time for maturing is eliminated unless the mixtures are slightly consolidated with a binder. High porosity enables good penetration and, similarly to lean mortars, their weak cohesion is well suited for observing consolidation effects. One of the best examples of this was reported by Price in 1984 [36], where a sophisticated procedure of impregnation is demonstrated. The preparation of crushed stone or compacted sand specimens is relatively straightforward. The first tests on the development of nanolime CaLoSiL® products were performed on specimens made of dry compacted mixtures. Such specimens are very fragile and must be transported and treated with care. The most important planning decision concerns the degree of packing of the dry mixture. The contents of the mixture may vary from very fragile materials to stone-like materials if a combination of pressing and vibration is applied. However, for consolidant testing softer materials are preferred. For compression tests, standard cubes ($40 \times 40 \times 40$ mm^3) are usually produced. For flexural tests reduced beam sizes ($20 \times 20 \times 120$ mm^3) are common. It is also possible to manufacture thin plates of compacted sand or crushed stone for tension tests. For bending tests a method of 'prosthesisation' can be used [13].

In the course of prosthesisation, a short sample of material is supplemented symmetrically on both ends with two 'prostheses' to the required length. The length must satisfy Navier's assumption of linear stress distribution along the cross section when loaded in flexure. This assumption forms the basis for the investigation of the technical elasticity and strength of materials, and has been used in deriving the mathematical models required for test evaluation. If a short sample need to be used, the measured deformation would be significantly influenced by the contribution of the shearing force. Wood has been found to be a suitable material for mortar prosthesisation. It is sufficiently strong, light, inexpensive and can be glued easily (Fig. 2.35). The material to be investigated is placed at the centre of the test specimen. Prosthesisation technology needs to address the problems encountered by using glued joints between the 'prostheses' and the samples. To counter this issue, it is essential to use an adhesive of sufficient strength in combination with adequate specimen loading. In this respect, three-point flexure appears more favourable than four-point flexure, as it places less stress on the area of the glued joint even though the influence of shear under the load is still present. Nevertheless, the moduli of rupture loads are rather low for fragile materials and the shear effect is not such an issue. In our experiments, we have modified

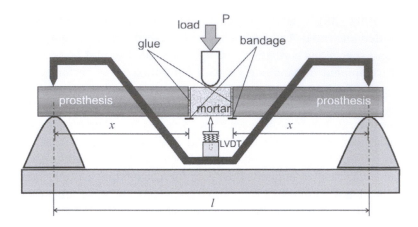

Figure 2.35 Three-point bending test of a mortar specimen prolonged with prostheses.

the method of attaching the wooden prosthesis to the mortar by 'bottoming' the glued joint (Fig. 2.36). When this measure was taken, the test specimens broke at the point of highest stress on the beam.

Thin stone plates are suitable for measurements of both thermal and hygric or hydric dilation. The dilation tests on mortar specimens require their casting into moulds and typically specimens of a cylindrical form are preferred.

Figure 2.36 Strengthening (bottoming) of the glued joint. Photo: M. Drdácký.

Vapour permeability is tested on the broken halves of the thin stone plates after flexural tests using specific purpose wet or dry cups. To perform the same tests on mortars, thin round plates are usually cast.

Other test specimens are designed for specific test purposes. Case studies involving these are presented in detail in Chapter 8.

2.3.3 Laboratory Tests of Consolidation Effects on Stone

The impact of consolidation treatment is investigated in a sequence of tests using the stone blocks and thin plates. Significant material changes are studied because they can have an important influence on the compatibility and durability of interventions. Material changes to the following factors are investigated:

- Penetration depth
- Thermal and hygric/hydric dilation
- Strengths and abrasion resistance
- Modulus of elasticity
- Vapour permeability
- Microstructure (porosity and pore-size distribution)
- Specific weight

When considering the consolidation of historic materials, a change of colour to the visible surface is also important. Water sorption rate and absorption, liquid contact angle, time for absorption and drying rate which must be tested on a larger bulk of material than is represented by a thin plate. The sequence of tests begins with non-destructive methods, that is, with colour measurements and ultrasonic assessment of the penetration depth and the dynamic modulus of elasticity. After cutting into thin plates, the next step is to measure the dilation characteristics of the material. Then the plates are loaded in three point bending up to failure in the laboratory environment typically of 20°C and 65% RH. During the course of loading, the deflection under the acting force is measured and from this data the modulus of elasticity is calculated. The broken parts are used for water permeability tests using a wet or dry cup method, and also for any other required physical measurements, including porosity. Such a sequence of tests ensures that all experimental data is generated from identical test specimens which improves the reliability of correlations and statistical calculations. Moreover the cutting method gives direct data on the consolidant penetration and any change its impact causes along the depth profile.

A specific approach had to be adopted when investigating hygric/hydric and temperature dilation coefficients. For such small test specimens an inexpensive method using a simple specific rig has been developed. Detecting and measuring minute changes in dimensions, even smaller than the surface and shape irregularities of the specimen, meant that the experiment had to be performed in a single sequence, without either handling or repositioning the specimen. The reason for this requirement becomes obvious when we consider that for example, sandstone specimens consist of grit held in a matrix, and the individual grains of grit are typically only 0.1 mm in diameter. However, the studied phenomenon – the dilation of the specimen – is in the range of a few micrometers. Therefore the order of difference between the two values makes it impossible to use anything other than a relative measurement of any length increase or decrease. In the case of the thermal dilation study, any changes in length of the small sandstone plates (about $15 \times 40 \times 3\,\text{mm}^3$) were measured while they were cooling from a preheated state to the ambient temperature of the laboratory or alternatively while they were drying from the saturated state to the natural moisture content corresponding to the temperature and relative humidity in the laboratory. In both measurements, the specimens are placed in the special rig and optical system shown in Fig. 2.37. During the thermal dilation tests, the changes in temperature are recorded using a thermocamera. The optical system contains a mirror which rotates in accordance with the variations in the length of the specimen and reflects a focused laser beam onto a screen. The movement of the laser beam spot on the screen is recorded using a digital camera with one pixel resolution (this is equal to a 0.06 mm shift of the spot).

Simultaneous recording of the position of the light spot and the temperature field in the specimen is vital for accurate measurement. In the evaluation procedure, the data from the image analysis must be studied alongside the data from the cooling (or drying) records. Any relationship is studied with the use of specially developed

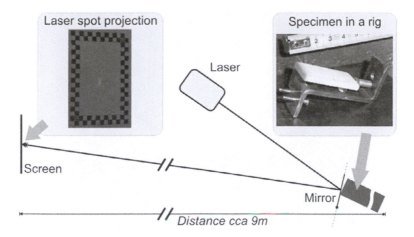

Figure 2.37 Schematic representation of the testing arrangement. The stone plate is placed on inclined supports and tends to slide down. Springs keep the mirror in contact with the specimen (not shown in the figure). Design: J. Valach.

software tools by means of time synchronisation. The position of the laser spot on the screen is determined as the centre of the light points and by their intensity. The temperature change (cooling) is usually recorded in intervals of 10 seconds over a period of ten minutes and the temperature changes vary from 20°C to 80°C. From the thermograms produced, the average temperature of the specimen is calculated, because during cooling a temperature gradient of several degrees is set-up in the inclined specimen. This data is then synchronised with the deformation measurements. For the hydric/hygric dilation, it is more efficient to use a system enabling measurements on several plates simultaneously with direct recording of the measured data. The plates of a typical length of 40–50 mm are inserted in a special rig (Fig. 2.38) and immersed in water or placed in an environment which allows a controlled change of relative humidity. The device even allows testing on much shorter plates (about 25 mm) when the specimens are available after the bending tests are complete. The change ΔL of the length is measured by means of LVDT sensors with the range up to 1 mm and the process of swelling time development is recorded in a data logger. Supposing the linear dependence of the moisture expansion on the absolute

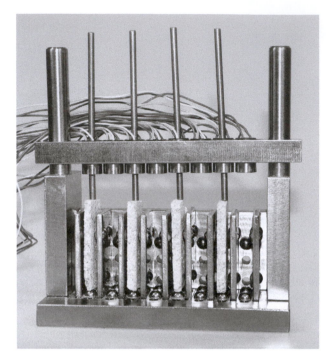

Figure 2.38 A frame for simultaneous measuring hydric/hygric dilation coefficients on up to eight thin plates. Photo: P. Zíma.

moisture content in the material enables to determine the coefficient of the moisture expansion α_H as a ratio between the change of the length ΔL at a change of moisture to the saturated value and the original length L as

$$\alpha_H = \frac{\Delta L}{L} \qquad (2.3)$$

The apparatus can easily give fluctuating measurements when the rig is being immersed in water and even small uplift forces can displace plates and influence the measured values. Therefore it is recommended to fix the plates to the base and then slowly fill the testing vessel from the bottom with water.

After the dilation tests, the thin stone plates may be tested in a destructive way. Bending strength and Young's modulus of elasticity are typically investigated on small-size specimens – rectangular

Figure 2.39 Arrangement of the thin plate three point bending test. Photo: M. Drdácký.

plates with nominal dimensions of $15 \times 40 - 50 \times 3.7 \, \text{mm}^3$. After maturing with a consolidant, the plates are cut in such a way that changes in the material properties along a depth profile perpendicular to the surface can be recorded. Sections of approximately 5 mm are practical, and give a plate thickness of about 3.7 mm with a cutting loss of about 1.3 mm. The specimens are tested in three point bending in a special rig after 24 hours of drying at 60°C and 12 hours of conditioning in a laboratory environment. During loading the deflection under the load is measured by a LVDT sensor and recorded for future calculation of the modulus of elasticity (Fig. 2.39). The typical test results are presented in Fig. 2.40 where an increase in the flexural strength (modulus of rupture) of the Maastricht limestone after nanosol CaLoSiL® E50 treatment is shown [37]. The full penetration depth even though

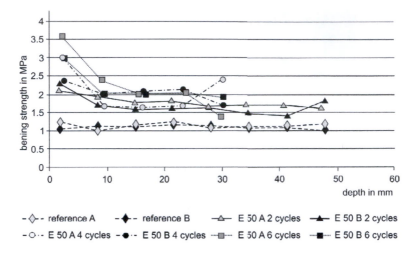

Figure 2.40 Bending strength of thin plates from Maastricht blocks – untreated (denoted as reference) and treated with nanosol CaLoSiL® E50 applied in 2, 4 and 6 cycles.

irregular and a gradient of the bending strength are clearly visible. The Maastricht stone is used as a suitable testing model material due to its low strength and high porosity. Penetration depth can be also tested using other mechanical characteristics, for example, cohesion manifested by abrasion resistance [38].

The broken halves of the thin plates are sufficiently large for the testing of vapour permeability, using a 'wet cup' method under controlled laboratory conditions (45% relative humidity, 25°C). In this method the loss of water (vapour) from a closed cup with 100% relative humidity passing through its cover 'membrane', created by using a thin stone square plate is measured. From this result, the vapour permeability is calculated using the weight loss recorded and the known cover plate area (Fig. 2.41). The dry cup version procedure can also be utilised.

Other non-destructive as well as destructive tests can be carried out on the broken parts of the thin plates. This sequence of tests reduces correlation errors generated by the diversity that occurs in a stone's material properties. Characteristics of stone which is rather non-homogeneous may vary widely within a rather small volume.

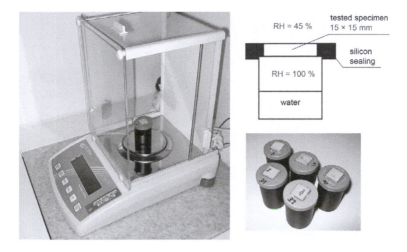

Figure 2.41   Dry cup vapour permeability test. Photos: M. Drdácký.

2.3.4 Laboratory Tests of Consolidation Effects on Mortar

The advantage of using thin walled specimens of stone for testing, allowed the development of similar methods for investigating mortars. This process was strongly influenced by a low efficiency of lime water impregnation test developed by Clifford Price in 1984 [36]. For example, tension tests are performed using thin plates (Fig. 2.42, left) while compression tests are carried out on short thin walled tubes (Fig. 2.42, right).

The plates for the tension tests are produced with wooden (plywood) heads to enable them to be fixed into the special flexible loading grips, ensuring the correct alignment without disturbing bending and without eccentricity (Fig. 2.42, left). All specimens are conditioned before testing in a controlled environment (20°C, RH 65%).

The short mortar tubes are loaded along the tube axis in compression tests. The compression strength results obtained using lean mortar specimens (lime : sand = 1 : 9) were checked against those measured on a set of rectangular specimens. The average compression strength determined from the untreated tubular reference specimens was 0.260 MPa, and the average compressive strength measured on the rectangular specimens was 0.549 MPa

Figure 2.42 Arrangement of the tension test (left) and a compression test (right). Photos: M. Drdácký, P. Zíma.

which, after applying a low slenderness ratio correction factor [13], corresponds to the cube compression strength of 0.365 MPa. In fact, it is not necessary to explore such a relationship for a comparative study of this type, because the effect of the individual consolidation agents is tested on identical material and the overall behaviour is compared. However, it follows from the results that the tubular specimens provide lower compression strength – approximately 70% of the values obtained using standard cubes, and that the poor mortar used is weaker than usual historic masonry lime mortars, for example [15]. It is worth noting that as this 'weak' mortar, was very well compacted and integrated and was not intended to model the typical sand-like disintegration of degraded mortars. The specimens of the lean lime mortar are typically cured by the slight spraying of distilled water onto their surface for about one month, which can have an adverse effect on the upper contact surface through which the compressive load is transmitted. This influence was studied in detail by means of printing the contact area using a simple 'frottage' technique and comparing the identified imperfections to the test

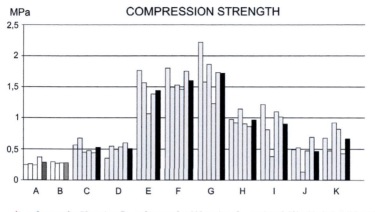

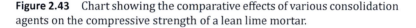

A – reference for 50 cycles, B – reference for 160 cycles, C – oxalate 2,5% with drying (38 cycles), D – lime water with drying (50 cycles), E – lime water with drying (161 cycles), F – CaLoSiL 25 (50cycles), G – CaLoSiL15 (40 cycles), H – CaLoSiL 15 (1 day application, 7 cycles), I – ethylsilicata NT 40 (2 cycles), J – silica (5 cycles), K –baryum water with drying (58 cycles)

Figure 2.43 Chart showing the comparative effects of various consolidation agents on the compressive strength of a lean lime mortar.

behaviour. On analysis of the test results the imperfections could be observed, where higher ultimate loads were usually attained on specimens with perfect contact surfaces. However, levelling the ends did not significantly improve the variation of the measured strengths in tests with lower ultimate loads. Only a slight increase in compressive strength was observed. Therefore, it is recommended that the specimens are tested with only one side levelled or without levelling at all. Nevertheless, for curing fresh lean lime mortar tubes a very gentle spraying is suggested.

The methodology of using thin walled specimens has been verified for other model materials including clay mortars. In relation to the nanolime this methodology was used for the comparative testing of effects of various agents on lime and clay mortars [35, 39]. One illustrative example is presented in Fig. 2.43 and Fig. 2.44.

The application of bending tests for the study of efficiency of consolidation agents does not generate such useful data but is possible if rather small specimens are produced, typically beams with $20 \times 20\,\text{mm}^2$ cross section and a minimum length of 120 mm.

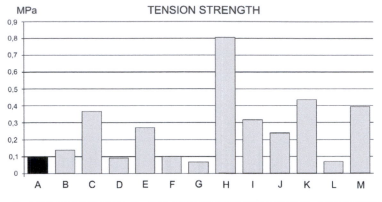

A – reference for 50 cycles, B – lime water with drying (119 cycles), C – baryum water with drying (53 cycles), D – sulphate water wet-to-wet (53 cycles), E – oxalate 5% with drying (53 cycles), F – oxalate 2,5% with drying (36 cycles), G – lime water + metakaolin (53 cycle), H – CaLoSiL 25 (40 cycles), I – CaLoSiL 15 (40 cycles), J – silica + CaLoSiL 25 (3 + 6 cycles), K – ethylsilicate NT40 (2 cycles), L – silica (5 cycles), M – silica + lime water (53 cycles)

Figure 2.44 Chart showing the comparative effects of various consolidation agents on the tensile strength of a lean lime mortar.

2.3.5 Laboratory Testing of Consolidation Effects on Compacted Sand or Crushed Stone

As previously mentioned, the compacted sand or crushed stone specimens are favoured as model materials. All types of tests can be carried out on such a material. However, the material is usually very fragile and it is practically difficult to manufacture beams. Therefore, the bending tests are typically carried out on regular long beams of the cross section of $20 \times 20\,\mathrm{mm}^2$ with a short central part of compacted sand elongated to the projected length with wooden prostheses glued to the both ends (Fig. 2.45, left). The untreated material frequently does not sustain handling and specimens would typically break before testing. Here the technique of prosthesisation may save such broken parts of the original specimen for further investigation. For compression tests cubes of standard mortar testing dimensions of $40 \times 40 \times 40\,\mathrm{mm}^3$ can be prepared (Fig. 2.45, middle). The compacted sand can also be tested in tension, typically using plates 10 mm thin and 40 mm wide (Fig. 2.45, right).

Compacted sand or crushed stone mixtures due to their good packing at low cohesion exhibit the most spectacular effects after

Figure 2.45 Practical arrangement for bending (a), compression (b) and tension (c) tests on specimens made of compacted sand or crushed stone. Photos: M. Drdácký, P. Zíma.

treatment, which proves the effectiveness of the consolidation agent. For example, test results on model specimens prepared from a compacted mixture of sand, marble powder and water demonstrated the excellent consolidation properties of the CaLoSiL® nanolime agents. The consolidation was shown to be effective when using compacted sand or crushed stone models, deteriorated stone or degraded mortars and plasters. The consolidation or strengthening effects observed are excellent especially taking into account the original very low strengths – in compression only 0.12 MPa, in bending and tension even in average 0.07 MPa. Using impregnation, the following increase in strength was measured. In compression testing with CaLoSiL® IP25 after 5 impregnation cycles an increase of 1717% was observed. After 10 impregnation cycles the increase attained was recorded at 3994%. CaLoSiL® E25 gave a 2875% increase after 5 impregnation cycles and after 10 impregnation cycles a rise of 4695% compared to the original value measured on the untreated material. The bending strength value recorded using CaLoSiL® IP25 after 5 impregnation cycles was up 507% and after 10 impregnation cycles showed an increase of 692%. Similarly, using CaLoSiL® E25 for 5 impregnation cycles, a 635% increase, and after 10 impregnation cycles a 1041% increase to the base strength. In tension testing the increase attained with the CaLoSiL® IP25 after 5 impregnation cycles was 1270% and after 10 impregnation cycles 2782%. Tension tests with CaLoSiL® E25 after 5 impregnation cycles showed an increase of 1348% and after 10 impregnation cycles 3025%. The results of mechanical tests show that the increase of the strength was higher on the specimen treated with CaLoSiL® E25 than on specimen treated with CaLoSiL® IP25.

2.3.6 In situ Testing of Consolidation Effects on Stone and Plaster

Non-destructive and partially destructive methods must be used when the consolidation efficacy is to be tested and evaluated on real objects after their treatment. It is usually required that any consolidation effect is demonstrated by quantitative measurement and therefore mechanical testing must be carried out. The in situ methods described in detail in Section 2.2 are applicable.

A specific technique using the fy Geotron have been developed during the STONECORE project for ultrasonic measurements in a depth profile (Fig. 2.46). The method consists of two probes – a US transmitter and a receiver – which are to be inserted into holes of 20 mm in diameter drilled into the surface layer to be investigated at

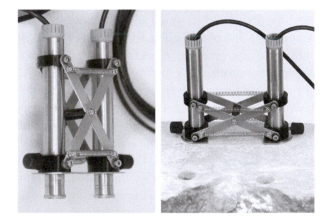

Figure 2.46 Portable ultrasonic double hole probe for the measurement of material properties along a depth profile. In addition the set-up is used for assessing the penetration depth at the surface of the material layer between two drilled holes. Photos: J. Valach.

a distance up to 100 mm and depth to 60 mm [40]. Features in design (a flat base, adjustable rods of the transmitter and the receiver) allow for the insertion of the device in prepared holes in order to get a reliable series of measurements for any changes occurring in the material investigated. The device is portable and fully compatible with US equipment for laboratory measurement. Therefore the operator can use skills previously acquired in data evaluation. The device is robust and well engineered and therefore it is suitable for measurements in the outside area. Thus it will allow the measurement of specific material properties in situ which had previously been very difficult. The technique also has little impact on the substrate being investigated. A study performed by J. Valach on sandstone demonstrated that this novel approach produced results comparable to those obtained by classical assessment methods (Fig. 2.47). The same study also showed a similar correlation between the two methods for other typical building materials (Fig. 2.48).

In addition, it was demonstrated that the new device can be used for the investigation of moisture in stone as there is a well-pronounced relationship between water content in stone and the change of the speed of the sound propagation (Fig. 2.49).

For the plaster cohesion assessment, the previously described peeling test seems to be the simplest method for checking the efficacy and quality of any near-surface consolidation.

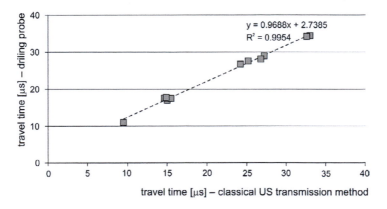

Figure 2.47 Comparison of the US wave travel time [μs] between classical transmission mode technique and the double hole probe. (The additive time constant – 2.7385 μs – is specific for an individual probe and can be easily subtracted from results to produce true data.)

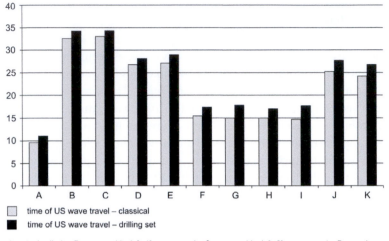

time of US wave travel – classical
time of US wave travel – drilling set

A – steel cylinder, B – spruce block [w1] – cross grain, C – spruce block [w2] – cross grain, D – marl
dry [01], E – marl wet [01], F – sandstone dry [SS1], G – sandstone wet [SS1], H – sandstone dry [SS2],
I – sandstone wet [SS2], J – brick dry [B1], K – brick wet [B1]

Figure 2.48 Comparison of the US wave travel time [μs] between classical transmission mode technique and the double hole probe for various materials. (The additive time constant – 2.7385 μs – from the previous figure explains the measured time differences.)

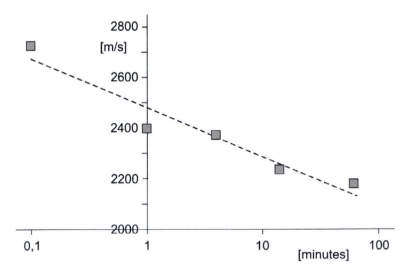

Figure 2.49 Speed of the sound wave in [m/s] as function of time after the initiation of wetting.

2.4 Microscopy as a Tool for the Characterisation of Materials*

2.4.1 Introduction

The term 'microscopy' stands for various magnifying imaging methods which are either based on a direct optical depiction of an object or its detail, or on the scanning of its surface with a focused beam, for example, of electrons. The most commonly used techniques in material sciences are light and scanning electron microscopy.

In general, microscopy provides more or less exact topographic information on a heterogeneous material, whilst the other physico-chemical techniques of analysis are capable to assess bulk information for a certain sample volume. Mineral materials such as stones or mortars represent such inhomogeneous solids, where a number of minerals of specific composition and morphology form a dense or porous microstructure which largely determines the material's macro-properties. Thus, microscopy, frequently by combining several techniques on just one sample preparation, offers excellent means to study and understand the properties of a given material in its sound and deteriorated states. In the field of conservation, this implies even the study of consolidants applied to a porous solid, namely in terms of the microstructure and the topographic distribution of the precipitate left behind in the pore space and bridging minerals grains once the liquid has evaporated.

In order to allow for a combined use of light and electron optical techniques in the required range of resolution, polished petrographic thin sections are usually the best choice of sample preparation. They can be taken for polarising microscopy in transmitted light – the classical approach to study stones in the geological sciences – as well as for incident light microscopy, before they may be taken to the scanning electron microscope for close-up observations and chemical spot analysis.

Petrographic thin sections are produced by cutting the material usually perpendicular to its surface, followed by gluing the plane

---

*This section is written by Johannes Weber.

onto a glass slide and grinding the section down to a thickness of approx. 25 μm. Porous solids require their impregnation by a resin as the first step of preparation, which is best done with the help of vacuum to force the resin even into narrow pores. It may be beneficial to colour the resin with a dye in order to enhance the visibility of voids, pores and cracks; to this end, often blue dyes are used since this colour rarely appears in the minerals. After grinding, additional polishing of the section can be of great value for the above mentioned reasons, even if this practice is still rarely followed by the laboratories.

2.4.2 Light Microscopy

Observations by polarising light microscopy (PLM) may provide a significant range of information on the mineralogical nature of the components and their textural and structural relationship.

In dependence of the fineness of the material to study, PLM inspections should start at comparably low magnifications to allow observing the basic microstructural properties including failure phenomena. Good stereo microscopes offer excellent possibilities for the low-resolution range. Various modes of illumination should be tried out at this stage, such as plain transmitted light or incident light at varying contrasts. When it comes to a close-up view on small details or when the optical properties of minerals need to be assessed for their identification, one may have to move the sample to a 'true' polarising microscope equipped with rotatable sample stage, polarisers for transmitted light observations, and additional incident light facilities.

Figures 2.50–2.55 show examples of thin-sections under the light microscope – both stereo and polarising. By a selection of some lithotypes and mortars characteristic of cultural heritage objects, some key properties relevant to their deterioration and conservation are illustrated and discussed in the captions. In turn, Figs. 2.56 and 2.57 refer to consolidants, in order to illustrate the sometimes limited potentials of light microscopy for conservation assessments. These limits are mainly due to the usual non- or low

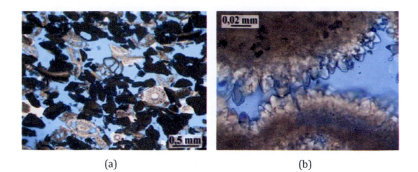

(a)   (b)

Figure 2.50 (a, b) Thin-section micrographs of a highly porous calcareous arenite, composed of microfossils and biogenic fragments. The pore space appears blue due to the impregnation with coloured resin prior to sectioning. At high magnification the calcite cement, a frequent product of diagenesis during the geological formation of the rock, can be observed as the mineral that keeps the components together. By chemical attack, for example, through air pollutants and water, this cement can be either dissolved or converted to sulphates. Both micrographs taken at plain transmitted light.

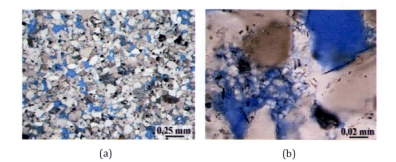

(a)   (b)

Figure 2.51 (a, b) Thin-section micrographs of a fine-grained silicate sandstone, composed of quartz and some feldspar. At high magnification, the kaolinite, a clay mineral derived from feldspar by diagenetic processes, can be observed as the matter that partly fills pores and yields cohesion of the quartz grains. Though not belonging to the typical swelling types of clay, kaolinite is quite sensitive to the action of moisture in the course of weathering. Both micrographs taken at plain transmitted light.

(a)   (b)

Figure 2.52   (a, b) Thin-section micrographs of a lean historic lime mortar with angular silicate rock fragments and a relatively high porosity. At higher magnification, the carbonate binder can be observed as typically fine-grained, that is, below the resolution of optical microscopy. Thus, the micropores existing between the binder crystals cannot be seen though they contribute to the well-known breathability of the mortar. Being an order of magnitude larger, the shrinkage cracks are well visible; they allow for capillary transport of solutions and provide the space for possible cycles of frost or salt crystallisation, or chemical attack of the calcite binder in the course of weathering. Both micrographs taken at plain transmitted light.

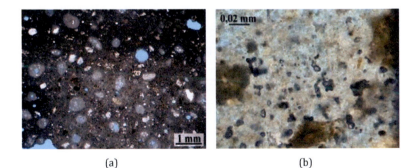

(a)   (b)

Figure 2.53   (a, b) Thin-section micrographs of a historic cement mortar; while at low magnification the fabric seems compact despite of a few isolated air voids, light bluish areas visible at high magnification indicate some amount of micropores in the matrix, probably a consequence of carbonation of the hydrate binder. Hardly accessible to liquid water, this pore system may yet allow for vapour permeability to some extent. Both micrographs taken at plain transmitted light.

Figure 2.54 Thin-section micrograph of an originally compact conglomerate with heavy decay due to weathering, revealed by a system of cracks (blue) and a thin black crust covering the surface. Position and size of cracks clearly indicate the places in need of consolidation and the consolidant's bridging capacities necessary to readjust cohesion. To allow the product to penetrate from the surface, the crust will have to be removed before the treatment. Plain transmitted light.

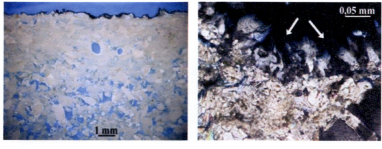

(a)   (b)

Figure 2.55 (a, b) Thin-section micrographs of a porous lime stone with decay due to weathering, indicated by a black crust on top of a several millimetres thick compacted zone (a). Close-up inspection (b) reveals the presence of gypsum (marked by arrows, growing into a pore) as product of reaction between calcite and sulphuric pollutants; directed migration of the solution followed by crystallisation of gypsum inside the stone pore system is the driving force of the compaction which may at some stage lead to the detachment of the surface. Any conservation treatment will face the risk not to penetrate through the compact zone. (a) incident light, (b) transmitted light at crossed polars.

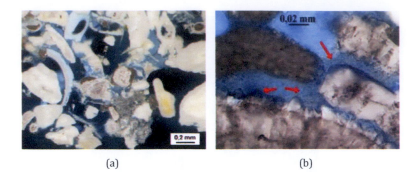

(a)   (b)

Figure 2.56   Thin-section micrographs of a porous limestone consolidated with nanolime. The precipitate can be seen by using various modes of illumination (marked by arrows), however an experienced eye is necessary to distinguish it from primary stone components. The bluish shade of the consolidant indicates its microporosity. In micrograph (b) the lime appears more grainy than in (a), an observation pointing to carbonation of the precursory calcium hydroxide. (a) transmitted light/dark field, (b) plain transmitted light. Photos: E. Mascha.

(a)   (b)

Figure 2.57   Thin-section micrographs of a porous limestone consolidated with tetraethyl orthosilicate (TEOS). Whilst at higher resolution (a) the precipitated silica gel can be recognised by its angular shape and some shrinkage cracks (marked by arrows), alternative modes may be required at lower resolution (b), beneficial to assess the in-depth distribution of the consolidant in larger diameters. For this specific consolidant, cathode luminescence microscopy [right-hand micrograph in (b)] proved powerful to visualise the silica gel by its greyish luminescence. (a) transmitted light, semi-crossed polarisers, (b) plain transmitted light (left) and cathode luminescence (right).

crystallinity of the new formed consolidating compounds which then lack birefringence as a primary feature of distinction under the polarising microscope. However, properties like, for example, the distinct cathode luminescence observed for some silica gels originating from ethyl silicate treatments, may yield interesting alternatives to PLM (Fig. 2.57b).

2.4.3 Scanning Electron Microscopy

Scanning electron microscopy (SEM) is a powerful tool to study a wide range of materials for their primary characteristics as well as for deterioration phenomena including secondary products of weathering and corrosion. Compared with optical microscopy, SEM can reach much higher resolution at excellent depth of focus. This compensates for the fact that no optical properties such as colour, refractive index or birefringence of minerals can be observed.

Three important effects occur when the electron beam in SEM is interacting with the sample surface, all of them are exploited by the study of inorganic mineral materials: first, secondary electrons (SE) of low energy are ejected from the very surface of the sample.

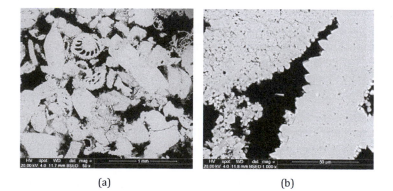

(a)   (b)

Figure 2.58 SEM-BSE micrographs of a porous calcareous arenite, composed of microfossils and biogenic fragments. Due to the monomineralic nature of the rock (just calcite as major constituent), the image is of a quasi-binary nature. The diagenetically formed calcite grain cement can be well recognised at higher resolution (b). Compare to Fig. 2.50.

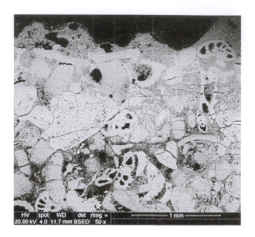

Figure 2.59 SEM-BSE micrograph of the same lithotype as Fig. 2.58, but from a surface zone compacted by secondary gypsum formed by the impact of sulphurous air pollutants. Compared to calcite, gypsum appears slightly darker under SEM-BSE conditions. Compare to Fig. 2.55.

The image contains morphological information based on 'light' and 'shadow' areas, thus this mode of observation is best for uneven planes as, for example, fractures. Second, back-scattered electrons (BSE) are reflected by the surface, the intensity of back-scattering being dependent on the mean atomic number and density of the sample area. In this way, compositional differences become visible which makes this mode of observation good for flat, polished surfaces. Third, X-ray emission is generated, its distinct bands of wavelength or energy being determined by the elements present in the sample. This enables qualitative or quantitative chemical analysis for a selected spot or area by energy-dispersive X-ray spectroscopy (EDX). The minimum diameter to obtain a clean analysis of a sample spot, however, is limited to about 5 μm.

Direct comparison with images from the optical microscope on thin sections makes use of the BSE detector combined with EDX. Apart from the possibility to achieve higher magnification in the SEM as compared to PLM, it also shows margins and boundaries between the grains in a clearer way, since the signals come right from the surface while in transmitted light the beam passes through the full

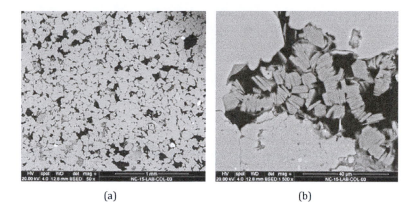

(a)   (b)

Figure 2.60 SEM-BSE micrographs of a fine-grained silicate sandstone (same sample as in Fig. 2.51), almost entirely composed of quartz. The diagenetically formed kaolinite can be well recognised at higher resolution (b); it occupies the spaces which are decisive for most weathering processes and consequently also for conservation treatments.

(a)   (b)

Figure 2.61 SEM-BSE micrographs of a historic lime mortar (same sample as in Fig. 2.52), composed of abundant rock fragments in a lime binder which reveals some minor shrinkage cracks. A close-up view of the binder (b) shows the presence of fines dispersed in the binder, an observation pointing to the use of unwashed sand in mortar preparation. Moreover, the high amount of micropores between the binder particles is visible, a feature characteristic of lime mortars which applied without much pressure. The specific places of vulnerability of lime mortars when exposed to moisture, frost and salts, however, can be found rather in the larger pores and shrinkage cracks (a) which thus may be in need of consolidation.

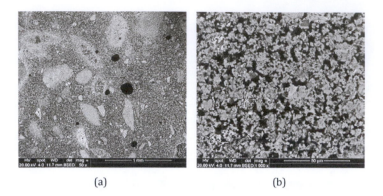

(a)   (b)

Figure 2.62 SEM-BSE micrographs of a historic cement mortar (same sample as in Fig. 2.53). At low magnification (a), no macropores despite a few air voids are visible; virtually all of the components represent unhydrated binder nodules, an observation characteristic of historic cements with their large temperature interval of calcination and coarse grinding. The resulting low rate of hydration, together with the carbonation of the hydrates at later stages, leads to the significant microporosity of the hydraulic binder matrix visible at higher magnifications (b) which comes close to the microporosity observed in lime binders (compare to Fig. 2.61 b). Thus, such cement mortars form strong, though fairly breathable systems which are even capable of some capillary moisture transport. If at all affected by weathering processes, they tend to develop cracks which may follow stress zones from early age shrinkage.

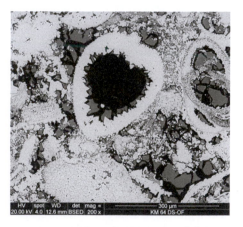

Figure 2.63 SEM-BSE micrograph of a calcareous arenite treated with TEOS. The silica gel deposited in the pores of the stone can be clearly distinguished from the calcite (bright) and the pores (black) by its dark grey colour. EDX spot analyses and mapping of Si can further confirm this observation.

Figure 2.64 SEM-BSE micrograph of a calcareous arenite treated with TEOS, studied on a sample fracture. The silica gel deposited between and covering the angular calcite crystals can be identified by its smooth and non-crystalline morphology and the shrinkage cracks.

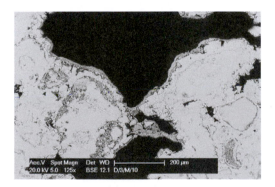

Figure 2.65 SEM-BSE micrograph of a porous limestone treated with a suspension of nanolime. Despite of the chemical similarity between substrate and consolidant, the latter can be well distinguished by its darker shade, due to a lower density of crystals packed in it. Photo: E. Mascha.

diameter of the section. In case of large areas to be investigated at low magnification, a certain disadvantage of the SEM technique lies in the fact that low magnification is restricted which necessitates to produce a series of micrographs which must then be stitched using an appropriate software.

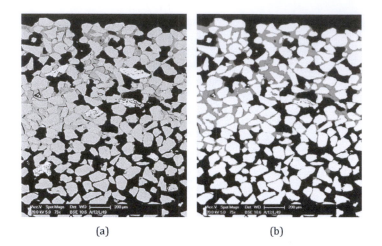

(a)   (b)

Figure 2.66   SEM-BSE micrograph of a laboratory aggregate sample from dolomite, treated with a suspension of nanolime from the surface on top of the image. The grey value of the consolidant in comparison to the substrate (a) is sufficiently distinct to allow for semi-automated pseudocolour editing (b), a procedure which not only enhances visibility but also allows for software-based calculation of quantitative parameters such as the rate of pore filling by the consolidant for selected layers of the specimen. In this way, the in-depth distribution of the precipitate can be studied at a resolution hardly achievable by other methods. Photos: E. Mascha.

2.4.4 Conclusion

Light and scanning electron microscopy offer powerful tools to study the microstructure of porous mineral materials in conservation as well as topography-related properties of conservation products applied to them. A combination of polarising microscopy and SEM at varying techniques of observation, performed on polished thin sections and sample fractures, respectively, yields the best results to the experienced eye. When cut perpendicular to the surface of atmospheric exposure and/or conservation treatment, a section observed by PLM and SEM reveals a high enough spatial resolution to account for microstructural and compositional gradients in depth, and hence to evaluate, understand and predict the most significant risks of failure for the material in its given state.

To link the insights in the microstructure to macro properties, assessed in non-destructive ways or on samples of the studied material, such as mechanical strength or petrophysical parameters, forms a major issue in conservation sciences. Due to the lower topographic resolution, or even bulk nature, of the latter approaches, however, such attempts are of limited success. It therefore appears that microscopy and other test methods should be applied in parallel, contributing each in its way to a better understanding on states and processes.

2.5 Chemical Composition, Chemical Reactivity and Their Determination*

2.5.1 Introduction

The determination of the mineralogical and chemical composition of the substrates to be conserved is an important precondition for the choice of suitable conservation strategies and materials. Whereas the characterisation of natural stone is generally realised by traditional mineralogical and petrographic methods including visual and microscopic methods, X-ray diffraction (XRD) and various spectroscopic techniques, detailed chemical analyses are needed to determine the salt content as well as the composition of historic mortars, renders or plasters. Additionally, the particle size has to be determined in order to characterise the aggregates. The complexity of this task becomes apparent by the fact that carbonates, for example, as well as silicates can be present in both binders and aggregates.

The presence of salts in stone, mortar and plaster may have several reasons. Possible sources are:

- Natural pollution (e.g., sea dust)
- The movement of groundwater into foundations
- The use of building materials containing salt during construction or restoration (e.g., clay containing sulphate)

---

*This section is written by Gerald Ziegenbalg.

- Environmental pollution (formation of sulphates, use of salt for de-icing)
- Conversion of organic matter (especially of urea into nitrates)

Salts can be present in a dissolved form in pore solutions or as crystalline minerals. Typical ions are Na^+, Ca^{2+}, Mg^{2+}, NH_4^+, K^+, SO_4^{2-}, Cl^- and NO_3^-. The corresponding salts are characterised by there high solubility in water, except gypsum ($CaSO_4 \cdot 2\,H_2O$), which is only slightly soluble. The presence of carbonate or hydrogen carbonate ions in the pore water can be ignored in the most cases.

Although the reactions and damage phenomena caused by salts are manifold and have been discussed in several publications [17, 41–43], a few fundamental principles, which are also of importance for any understanding the complexity of salt analyses, are summarised below.

The solubility of a component in water is constant at a given temperature. For example, the solubility of NaCl in 100 g water at 25°C is 35.88 g. The solution is saturated in NaCl, which means that the solution is unable to dissolve any more salt. The saturation concentration can be expressed in weight or volume-based units. These can be converted into each other by the solution density. The solubility of a salt mineral depends on temperature and is influenced greatly by the presence of other soluble components. Whereas temperature changes have little effect on the solubility of NaCl, a significance temperature-related increase can be observed for KCl or KNO_3, for example. A completely different behaviour is typical for anhydrite ($CaSO_4$) – the solubility decreases with the temperature (see Fig. 2.67, based on the data given in Ref. [44]).

Many salts form different hydrates. Sodium sulphate, for example, may occur as $Na_2SO_4 \cdot 10\,H_2O$ (Glauber's salt) or water-free Na_2SO_4 (Thenardite). The transition point, that is, the point where the two salts occur together, is 32.4°C. $Na_2SO_4 \cdot 10\,H_2O$ is the stable solid at lower temperature and Na_2SO_4 at higher temperatures. Both salts differ completely in their solubility behaviour. $Na_2SO_4 \cdot 10\,H_2O$ is characterised by an increase in solubility when the temperature rises up to 32.4°C. The solubility of Na_2SO_4 decreases with increasing temperature (see Fig. 2.68, according to Ref. [45]).

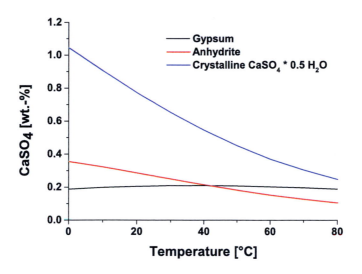

Figure 2.67   Solubility of different CaSO$_4$ minerals.

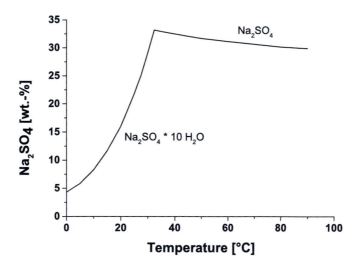

Figure 2.68   Solubility of Na$_2$SO$_4$. The solutions are unsaturated below the solubility curve.

To induce crystallisation processes, the solution has to become supersaturated, in other words slightly more salt has to be dissolved than is possible to achieve an equilibrium. Supersaturated solutions can be produced by temperature changes or by evaporation of the solvent (water). Supersaturation is the driving force behind crystallisation. The higher the level of supersaturation, the faster crystallisation will occur. Crystals with a small particle size are formed typically. In contrast, slow crystallisation rates and the formation of large crystals are typical for solutions characterised by low supersaturations. The overall influence of pressure on the solubility is slight compared to temperature.

Typical salts occurring in masonry, mortar or plaster are:

- $CaSO_4 \cdot 2\,H_2O$: Gypsum
- $MgSO_4 \cdot 7\,H_2O$: Epsomite
- Na_2SO_4: Thenardite
- $Na_2SO_4 \cdot 10\,H_2O$: Glauber's salt
- $3\,CaO \cdot Al_2O_3 \cdot 3\,CaSO_4 \cdot 32\,H_2O$: Ettringite
- $CaCl_2 \cdot 6\,H_2O$: Calcium chloride hexahydrate
- $NaCl$: Halite
- $Mg(NO_3)_2 \cdot 6\,H_2O$: Magnesium nitrate hexahydrate (magnesia saltpetre)
- $Ca(NO_3)_2 \cdot 4\,H_2O$: Calcium nitrate tetra-hydrate (Nitrocalcite)
- $5\,Ca(NO_3)_2 \cdot 4\,NH_4NO_3 \cdot 10\,H_2O$: Calcium-ammonium-nitrate
- $Na_2CO_3 \cdot 10\,H_2O$: Sodium carbonate decahydrate (washing soda, natron)
- K_2CO_3: Potassium carbonate

Whereas gypsum and magnesium sulphates are often the result of environmental pollution by sulphur dioxide (SO_2) and a subsequent reaction with calcium or magnesium carbonates, nitrates are normally the result of organic pollution and are formed by biological urea decomposition. Alkali carbonates may be the result of previous treatments with alkali-silicate solutions.

The moisture content of stone, plaster and mortar is often the subject of controversial discussions. A distinction has to be made between two principally different sources of moisture:

Table 2.3   Water vapour pressure of saturated salt solutions

Vapour Pressure	Temperature [$°C$]			
[Torr]	**NaCl**	**KCl**	**MgCl$_2$**	**KNO$_3$**
20	26.8	24.5	41.7	24.8
30	34.2	32.3	50.8	30.6
50	43.7	41.8	62.3	39.3
100	57.5	55.4	79.5	55.3

- Moisture present in stone, mortar or plaster resulting from the specific properties of the material (porosity, capillarity, condensation) and the environmental conditions (relative humidity, rain, groundwater, etc.)
- Moisture bound in hygroscopic salts or as hydrates

Hygroscopicity is the ability of substances, especially of salts, to bind moisture from the environment. All hygroscopic salts are characterised by an excellent water solubility. This means that, large amounts of salt can be dissolved in small volumes of water.

Saturated solutions always constitute the equilibrium between solids and a solution. If the humidity in a room is higher than the vapour pressure of the saturated solution, the solution will absorb moisture and further salts can be dissolved. Conversely, drying occurs in rooms characterised by a lower humidity and additional salts will crystallise. Some typical vapour pressures of saturated solutions for different salts are shown in Table 2.3.

Similarly, temperature changes result in dissolution and crystallisation cycles, depending on the solubility characteristics of the salt.

2.5.2 Determination of the Salt Content of Stone, Mortar and Plaster

When salt efflorescence, meaning the appearance of salt crystals on the surface, is observed, sampling is normally not a problem and sufficient material for analyses can be obtained. The characterisation of the salt content of mortar or stone proves to be more difficult. Collecting representative samples always poses a challenge due to their heterogeneous nature. Moreover, often only small samples can be taken so that the amount available for analyses is limited. This has to be taken into account when evaluating the analytical results with respect to possible conservation strategies.

The traditional method to determine the salt content is based on leaching the sample with water followed by chemical analysis using standard analytical methods.

In a first step, the sample (when necessary roughly milled) is dried at 120°C until its weight remains constant. It should be noted that the weight lost is the humidity and water that has been released from salt hydrates.

In some cases, 120°C are not sufficient to de-water all hydrates completely, but this temperature is seen as favourable in most cases. Drying at lower temperatures is possible, but may result in only partial de-watering of the existing salt hydrates.

In the next step, homogenisation of the sample by milling is of great importance to guarantee a quantitative leaching of the salts, especially when materials with small pore radii are to be characterised. Milling should be carried out to a size <120 μm.

The volume of water used to extract the solids is extremely important for the quality of the resulting analyses. If the salt content is low and the volume of water is too high, the analyses may come up against the detection limits of some analytical methods. Most of the salt minerals that occur in stone and mortar are characterised by a high solubility, leaching stone and mortar samples will not usually result in the formation of saturated solutions. Water-to-solid ratios that do not allow a complete leaching of all salts are not to be expected in most cases. For example: if sodium sulphate is present in the samples, leaching 100 g of stone with 250 g of water would require the presence of 45 g Na_2SO_4 in order to produce a sodium sulphate (Na_2SO_4) saturated solution at 20°C. Similar calculations for other salts/salt hydrates show that the solubility limits can only be exceeded, in the presence of abnormal, extremely high salt loads. One exception is gypsum, which has a solubility of approximately 2.5 g/L in water at 25°C. If near-surface zones or crusts formed on weathered stone or mortar in particular are to be analysed, high gypsum contents are possible. This means that, leaching with low volumes of water may produce data that does not reflect the total gypsum content. In order to avoid this, the analytical data obtained has to be checked thoroughly, especially with respect to the Ca^{2+} to SO_4^{2-} ratio. If the data indicates that the gypsum solubility has been exceeded, repeated analyses of samples obtained by leaching with higher water contents are necessary. Another characteristic of gypsum is that its dissolution rate in water is much slower than that of the other salt minerals. If crusts containing dust and biological material are to be characterised, dissolution is slow and calls for an adequate extraction time. The leaching step should always be carried out in stirred solutions.

Commonly used analytical techniques are ICP-AES (inductively coupled plasma atomic emission spectroscopy), AAS (atomic absorption spectroscopy), ion chromatography, photometry as well as traditional methods such as titration or gravimetry. In addition to the chemical analyses, the pH value and conductivity of the leachate should be determined. The following methods are seen as favourable with and adequate sample size:

- Na^+, K^+: Determination by atomic absorption spectroscopy
- Ca^{2+}, Mg^{2+}: Complexometric titration with EDTA (ethylene-diamine tetra-acetate)
- Cl^-: Potentiometric titration with $AgNO_3$
- NO_3^-: Photometric determination
- SO_4^{2-}: Gravimetric determination as $BaSO_4$

A particular problem is that one salt dominates in many cases and strong dilution is necessary for its quantitative determination. If the entire sample is diluted, this may result in concentrations of the other components that are difficult to detect. Thus, it is always very important that sufficient solution volumes are available for different dilutions and analytical methods.

All necessary dilutions have to be carried out carefully. They are often the source of significant errors. It has to be remembered that the use of extremely high dilution factors is always critical, especially when high salt loads are expected. Using pipettes with larger volumes and gradual dilution can reduce the errors significantly. The pipettes used for dilution have to be calibrated periodically, for example by weighing the volume of water.

In order to assess the quality of the analyses, the total sum of cations and anions has to be calculated to obtain an ion balance. The total anion concentration should be equal to the total concentration of cations. The resulting ratio of total anions to cations should be equal to one. The concentrations must be expressed in terms of ion equivalents (mEq: milli-equivalent), which can be calculated using Eq. 2.4:

$$mEq = \frac{(c \cdot v)}{M} \tag{2.4}$$

c: concentration in mg/L; v: valency of the ion; M: molecular weight
    The analytical error is shown in Eq. 2.5:

$$Error = \frac{\left(\sum cations - \left|\sum anions\right|\right) \cdot 100}{\left(\sum cations + \left|\sum anions\right|\right) \cdot 0.5} \tag{2.5}$$

The following analytical errors are acceptable.

- Sum of cations and anions, respectively $> 1\,mmol/L : 5\%$
- Sum of cations and anions, respectively
  $< 1\,mmol/L : \leq 0.1\,mmol/L$

An initial overview of the salt content and necessary actions is obtained by analysing only the anions chloride, nitrate and sulphate. The assessment of different contents is shown in Table 2.4 [46].

Table 2.4   Assessment of salt contents in masonry

	Concentration [wt%]		
Chloride	<0.2	0.2–0.5	>0.5
Nitrate	<0.1	0.1–0.3	>0.3
Sulphate	<0.5	0.5–1.5	>1.5
Assessment	Low contamination; treatments are only necessary under special circumstances	Average contamination; exact determination of the salt content is required	High contamination; exact determination of the salt content is required; treatment required for salt removal

2.5.3 Quantitative Mortar Analyses

The redevelopment of historic mortars requires a knowledge of the binder used, the binder-aggregate ratio as well as the particle size distribution of the aggregates. Furthermore, the origin of the aggregates is also of interest. The analytical characterisation of historic mortars is always a challenging task and often difficult, if not impossible. The main reasons for this are the decomposition of the original natural organic materials that were used and possibly a lack of knowledge of the conditions during binder production. For example, the reactivity and properties of air lime-based binders are greatly influenced by the composition of the limestone used as well as the burning and slaking conditions. It will often be impossible to reconstruct the technology used, for both technical and economic reasons. Thus, the main challenge is to produce mortars based on materials that are available today, but whose properties are as similar as possible to those used originally.

X-ray diffraction (XRD) measurements and thermoanalytical investigations are extremely helpful to gain an initial overview of the binder and aggregate composition. Whereas the first method provides qualitative information about the phase composition, thermoanalytical measurements allow a quantitative determination, mainly of gypsum, calcium carbonate and calcium hydroxide. In addition, information is obtained about salt hydrates that may be present.

The first step is to separate binder and aggregates. Although this is difficult, and generally incomplete, fairly good results are achieved if carefully ground materials are sieved into fractions above and below 90 μm. Before this, the samples have to be dried at 40°C. It is assumed that the fraction >90 μm represents aggregates, while particles <90 μm are attributed to the binder.

Thermal analyses such as thermogravimetric analysis (TGA) and differential thermal analysis (DTA) are techniques whereby a sample is heated at a constant heating rate in a defined atmosphere. The weight and temperature of the sample in question and of an inert sample (Corund, Al_2O_3) are recorded during heating. The differences between the two samples, both in weight and temperature, indicate thermal decomposition processes or phase changes. Figure 2.69 shows a typical set-up that is used for thermoanalytical measurements.

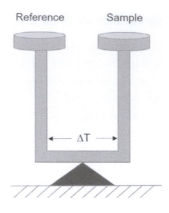

Figure 2.69 Typical set-up for DTA measurements. Sample holder with two thermoelements for the inert reference sample and the sample of interest. Determination of the differences in temperature between the probes and weight changes during heating.

Weight losses that appear in the TG graph indicate thermal decomposition, for example de-watering or decomposition reactions of hydroxides or carbonates. The corresponding DTA signals characterise the decomposition temperatures. DTA signals without corresponding TG signals indicate phase transitions (e.g., melting, modification changes, etc.).

The DTA peak characterises the heat associated with the phase transition/decomposition reaction. An exact quantitative analysis requires calibration of the system. In most cases, the peak temperatures and the weight changes are sufficient to identify the components. The following reactions, which allow an exact identification of present phases, are important for mortar and plaster analyses:

- 100°C–150°C: $CaSO_4 \cdot 2\,H_2O \rightarrow CaSO_4 + 2\,H_2O$
- 400°C–550°C: $Ca(OH)_2 \rightarrow CaO + H_2O$
- 450°C–550°C: $MgCO_3 \rightarrow MgO + CO_2$
- 650°C–900°C: $CaCO_3 \rightarrow CaO + CO_2$

Since only small sample amounts are used (10–20 mg), homogenisation of the starting material is of great importance. The standard heating rate is 10 K/min, an inert gas flow (nitrogen) should be used

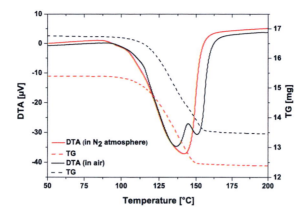

Figure 2.70 DTA/TG results for $CaSO_4 \cdot 2\,H_2O$. Black line: heating rate 5 K/min measured in air. Red line: heating rate 5 K/min, in N_2 atmosphere (100 mL/min).

at rates typical for the analytical system (often 100 mL/min). Typical DTA/TG graphs are shown in Figs. 2.70–2.72.

The mortars can be classified on the basis of the results of XRD and/or thermoanalytical investigations. Both methods provide important information for further analyses that may be necessary.

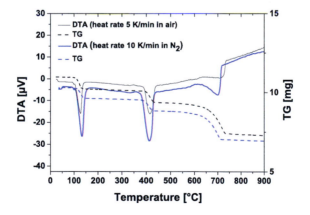

Figure 2.71 DTA/TG results for a mixture of $Ca(OH)_2$, $CaCO_3$ and $CaSO_4 \cdot 2\,H_2O$. Blue line: heating rate 10 K/min and constant nitrogen flow (100 mL/min). Black line: heating rate 5 K/min in air.

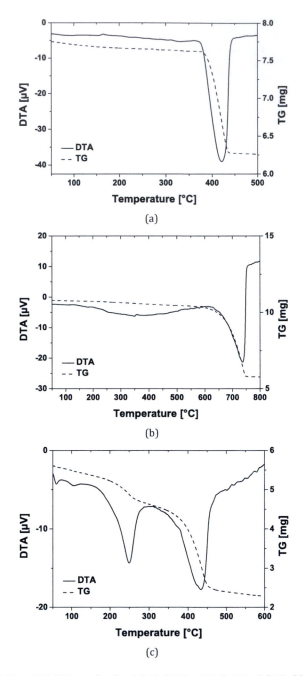

Figure 2.72 DTA/TG results for (a) Ca(OH)$_2$; (b) CaCO$_3$; (c) MgCO$_3$ with a heating rate of 10 K/min and constant nitrogen flow (100 mL/min).

The use of chemical analyses to characterise mortars and plasters has been the subject of numerous discussions over the years [47]. Before presenting and discussing a possible methodology, it should be mentioned that chemical analyses can not answer every question, especially when similar compounds are present in the binder and aggregate fraction, which is quite typical for calcareous systems.

Several ways in which selective leaching techniques can be used to separate calcareous binders and aggregates have been discussed, such as leaching with salicylic acid or by solutions containing EDTA as complex forming agent. These proposals may be applicable in special situations, but it is also well known that often only partial separation is possible.

One frequently discussed method to determine the binder-aggregate ratio is the acidic dissolution of the sample. The idea behind this is based on the assumption that the calcareous binders will dissolve while the aggregates remain unaffected. This is a favourable method, when lime mortars with siliceous aggregates have been used. If carbonate-bound aggregates are present, the method quickly reaches its limitations. The presence of any pozzolanic materials, such as natural hydraulic lime, can also cause significant problems.

The following analytical method is convenient to characterise simple calcareous mortars containing siliceous aggregates.

- Drying at 120°C until a constant weight is achieved. Drying at lower temperatures, for example at 40°C, has often been suggested. However, it takes a long time until constant weights are achieved and there is always the risk that not all of the water will be released.
- Manual separation and careful grinding (without destroying the aggregates). Sieving: <90 μm = fraction rich in binders; >90 μm = aggregates

  The particle size distribution can be determined by sieving or laser diffraction measurements. If laser diffraction is used, it is very important that the measurement principles are understood and the evaluation parameters adjusted to the characteristics of the sample in question (e.g., the refractive index of the minerals).

- Leaching of the fraction rich in binders ($<90\ \mu m$) with water followed by filtration

  - Filtrate: determination of the pH value, conductivity, Na^+, K^+, Mg^{2+}, Ca^{2+}, SO_4^{2-}, Cl^-, NH_4^+, NO_3^- ($=$ soluble salt content)
  - Residue: drying $+$ weighing $=$ insoluble residue in water

- Acid dissolution (HCl) of the insoluble residue (from 3)

  - Filtrate: determination of Ca^{2+}, Mg^{2+} ($=$ characterisation of present carbonates) and, if necessary of other components such as Si^{4+}, Fe_{total} or Al^{3+}
  - Residue: washing, drying and weighing $=$ insoluble acid residue $=$ siliceous components in the mortar

- Leaching with hot Na_2CO_3 solution: soluble silica of the binder ($=$ hydraulic components, C-S-H phases)

The following remarks are necessary:

- Grinding will never achieve a complete separation of aggregates and binders. Thus, it should always be remembered that each fraction contains parts of the other.
- Air lime-based mortars often contain unreacted calcium hydroxide also after longer periods of times. High pH values of the leach solution (step 3) indicate the presence of $Ca(OH)_2$.
- Measuring the conductivity of the aqueous leachate provides an initial overview of the presence of soluble salts. The following formal conversion can be used to calculate the total salt content: $1\ mS = 640\ mg\ KCl/L$.
- The amount of binder that has to be brought into contact with hydrochloric acid as well as the concentration of the hydrochloric acid depends on the goal of the analysis and the reactivity of the mortar. The most favourable amount has to be determined by experimental tests. In general, the amount of binder used should allow a safe analysis of all components of interest in concentrations that are far above the detection limits. The HCl concentration should always be in stoichiometric excess with respect to the expected $CaCO_3$ content.

- Leaching the acid-insoluble residue with hot sodium carbonate solution results in the amount of soluble silica. This reflects the hydraulic components (hydraulic lime, etc.) content in the binder.
- All filtration and washing steps should be carried out until the filtrate is free from chloride. To prevent an unreasonable increase in volume, repeated washing with small amounts of deionised water is favoured.
- As has already been mentioned for salt analyses, all values have to be checked carefully with respect to their probability.

The procedure can also be used to characterise the composition of calcareous aggregates. When analysing gypsum-based plasters, the same procedure can be used. It should be remembered, however, that larger volumes of water and higher concentrations of hydrochloric acid are required due to the low solubility of gypsum in water. A detailed description of one possible analytical method can be found in Ref. [47].

Chapter 3

Inorganic Binders and Consolidants: A Critical Review

Gerald Ziegenbalg,[a] Zuzana Slížková,[b] and Radek Ševčík[b]

[a] *IBZ-Salzchemie GmbH & Co. KG, Schwarze Kiefern 4, 09633 Halsbrücke, Germany*
[b] *Institute of Theoretical and Applied Mechanics of the Czech Academy of Sciences, Prosecká 76, Prague 9, Czech Republic*
gerald.ziegenbalg@ibz-freiberg.de; slizkova@itam.cas.cz; sevcik@itam.cas.cz

3.1 Introduction

In many parts of the world, historical as well as modern buildings, sculptures and monuments are based mainly on inorganic natural or artificial components such as silicates, carbonates or sulphates. The variety and combination of such minerals and components is immense, ranging from natural stones such as sandstone, granite, marble and limestone to complex composite mortars, plasters and renders, as well as artificial stone such as concrete.

In most cases, combinations of different materials are found, for example lime mortar connecting bricks or render protecting masonry. Each material and material combination has its own, specific properties and damage phenomena.

Nanomaterials in Architecture and Art Conservation
Gerald Ziegenbalg, Miloš Drdácký, Claudia Dietze, and Dirk Schuch
Copyright © 2018 Pan Stanford Publishing Pte. Ltd.
ISBN 978-981-4800-26-6 (Hardback), 978-0-429-42875-3 (eBook)
www.panstanford.com

For centuries, the only traditional materials available for the conservation of stone, mortars and renders were mainly lime, gypsum, linseed oil, wax, animal and vegetable glues. Later, with the development of chemistry, components such as water glass, fluorosilicate, barium hydroxide, ethyl-silicate and synthetic resins entered into building materials conservation practice [17]. Many of these chemical substances failed to satisfy the compatibility and performance requirements for consolidation treatment. The use of organic polymers (acrylates, acrylamides, urethanes, epoxides, etc.) requires careful discussion, especially with respect to compatibility and long-term stability. Many examples are known where inappropriate combinations or weathering of polymers [48–50] (especially by UV radiation) have resulted in serious damage. Organic compounds will not be further discussed in the following.

One of the most successful groups of stone consolidants in use are silicic acid esters (SAEs). These will be discussed in detail in Chapter 6, which describes the combined application of silicic acid esters and calcium hydroxide dispersions.

Inorganic binders and consolidants can be divided into systems containing:

- Reactive solids dispersed in aqueous suspensions
- Reactive aqueous solutions containing only dissolved inorganic components
- Nanoparticles suspended in different solvents

The first group contains binders that solidify over time and form mechanically stable masses. Cementitious materials and systems based on reactive calcium hydroxide suspensions as well as suspensions leading to the formation of gypsum on its own or in combination with other components are typical for this group. Because this book focuses on systems based on calcium hydroxide, this group will be discussed in detail. Excellent summaries exist elsewhere that describe cementitious systems [51, 52] as well as gypsum-forming masses [53, 54]. A discussion of these groups would go beyond the constraints of this book. Because of the relatively large size of particles in suspensions, their penetration into the porous structures is limited. These binders are mainly used for steps that follow or are connected with the structural consolidation such as grouting or repairs to surface cracks and defects.

The second group, reactive aqueous solutions, affords a much deeper penetration into porous or fractured substrates than suspensions due to the character of the solutions. They form amorphous gels or react with the materials to be treated to form secondary minerals consisting of phosphates, oxalates or sulphates.

Over the past few years, the third group, suspensions or dispersions of nanoparticles, have been investigated as novel consolidant agents. Nanoparticles can be suspended in different solvents such as ethanol and may offer several advantages compared to those consolidant agents already in use. The main aspects of the use of nanolime will be discussed in detail in subsequent chapters. A comprehensive survey of recent consolidation materials, laboratory test methods and in situ tests on render/plaster before and after consolidation can be found in Ref. [55].

3.2 Characteristics of Aqueous Calcium Hydroxide Suspensions

Before discussing selected materials in detail, a definition of the available classes of calcium hydroxide-based materials and in particular a discussion of the terms used would appear necessary. This is of great importance because incorrect names, old trivial names and trademarks may cause confusion and an incorrect choice of conservation materials. In the following, the terms that can be found in DIN EN 459-1: 2015, including the following definition of lime, will be used here:

- Calcium oxide and/or calcium magnesium oxide obtained by thermal decomposition (calcination) of naturally occurring calcium carbonate (limestone, chalk or shell limestone, as well as of dolomitic limestone or dolomite)
- Calcium hydroxide and/or calcium magnesium hydroxide

The mineralogical name of calcium hydroxide ($Ca(OH)_2$), which has a hexagonal crystal lattice, is portlandite. Because lime has many applications with different requirements, DIN EN 459-1:2015 provides the following special definition for building lime:

'Group of lime products, exclusively consisting of the two families; air lime and lime with hydraulic properties, used in applications or materials of construction, building and civil engineering.'

3.2.1 Air Lime

Air lime has no hydraulic properties and is defined as lime which combines and hardens by reaction with atmospheric carbon dioxide. Air lime can be split into the subfamilies:

- Calcium lime (or white lime)
- Dolomitic lime

Forms of air lime are:

- Caustic lime, quick lime (air lime consisting of oxides that react with water in an exothermic reaction)
- Lime hydrate, slaked lime (air lime, mainly consisting of hydroxides, which was obtained by slaking the oxides in water)

For centuries, calcium oxide (CaO, lime) has been obtained by burning limestone at temperatures above 900°C. It is an endothermic, energy-consuming reaction. Limestone was traditionally burned in uncovered heaps consisting of alternating layers of wood and limestone. Today, different types of shaft and rotary kilns are used. The properties of the resulting calcium oxide are largely influenced by several factors, important ones being:

- The composition and characteristics of the limestone used as well as its particle size distribution
- The conditions of the thermal treatment, such as absolute temperature and temperature profile during the steps heating, calcination and cooling
- The retention time and the CO_2 content of the atmosphere during calcination

The final product air lime is obtained by the reaction of calcium oxide with water.

$$CaO + H_2O \longrightarrow Ca(OH)_2$$

The conversion of calcium oxide into calcium hydroxide (air lime) is a strongly exothermic reaction, the heat generated during the hydration of 1 kg CaO is sufficient to heat 2.8 kg H_2O from 0°C to 100°C. This so-called slaking can be realised in two ways:

(1) Water is applied in excess to the amount required by stoichiometry (0.321 kg H_2O/kg CaO). The process is often called wet-slaking and is carried out in lime pits. Coarse as well as fine calcium oxide particles are treated with water while the suspension is being stirred. Due to the exothermic character of this reaction, the suspension heats up. Mixing has to be carried out under conditions that prevent the suspension from boiling. The most common water : calcium oxide ratio is at around 3 : 1. After slaking, the resulting calcium hydroxide suspension is stored in the lime pit or in special containers to allow an ageing of the calcium hydroxide particles. The particles undergo significant changes to their morphology during this process. Prisms that formed originally turn into plate-like crystals. This process is accompanied by a reduction of the particle size due to:

- Differences in the solubility between {0001} basal pinacoid faces and {1010} prism faces [56]
- Heterogeneous secondary nucleation on a nanometerscale, plate-like portlandite crystals on pre-existing larger $Ca(OH)_2$ crystals [57]

Both processes result in improved workability, plasticity and water retention. In building practice, the matured calcium hydroxide suspension is commonly called lime putty.

(2) In contrast to wet slaking, which results in a suspension as a final product, dry slaking produces powders. In order to achieve this, water is only added in amounts that are sufficient to convert calcium oxide into calcium hydroxide and to compensate the amount of water that is evaporating. Calcium oxide particles with well-defined particle sizes are brought into contact with hot water under intensive mixing conditions. The final moisture content of the calcium hydroxide powder should not exceed 1 wt%.

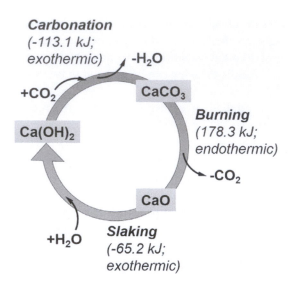

Figure 3.1 The lime cycle.

The so-called 'lime cycle', the conversion of calcium carbonate (limestone) via burning, slaking and subsequent carbonation back into calcium carbonate is illustrated in Fig. 3.1.

The final step in the lime cycle, the carbonation of calcium hydroxide through the formation of calcium carbonate, is a very complex, multistep process consisting of the dissolution of gaseous carbon dioxide in water, dissociation reactions and final precipitation/crystallisation steps. A detailed overview of possible reaction mechanisms will be provided in Chapter 4 along with a discussion of the carbonation of calcium hydroxide nanoparticles. For the present, we shall only focus on the different varieties of calcium carbonates that can be produced (see Fig. 3.2.1). A localised quantification of anhydrous calcium carbonate polymorphs using micro-Raman spectroscopy was recently published [58a]. The following polymorphs are known:

Calcite: Thermodynamically the most stable $CaCO_3$ modification under earth surface conditions, with a trigonal crystal system (hexagonal scalenohedral crystal class). It is formed by crystallisation from aqueous solutions and biomineralisation.

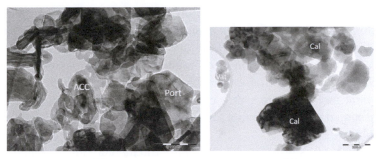

(a)   (b)

(c)

Figure 3.2 TEM of different $CaCO_3$ modifications formed during the carbonation of nanolime (CaLoSiL®) (a) ACC and unreacted portlandite (Port) after 6 h, (b, c) Complete carbonation after 24 days showing calcite (Cal), vaterite (Vat) and aragonite (Ara). Photos: Carlos Rodriguez-Navarro.

Vaterite: The crystal structure of vaterite is still a matter of scientific interest. Vaterite's structure most probably occurred as a combination of different forms [59, 60]. It can be converted into aragonite or calcite and is found, for example, in kidney and gall stones and plays a significant role in biomineralisation processes. Vaterite has also been discovered in historical mortars (see Refs. [61–63]).

Aragonite: Metastable modification with orthorhombic crystal system (dipyramidal crystal class). Conversion into calcite takes place. It is found in natural pearls, corals, etc.

Amorphous calcium carbonate (ACC): Normally hydrated, often with a composition similar to that of $CaCO_3 \cdot 1... 1.5\,H_2O$ [64]. The structure of ACC is based on a nanoporous framework rich in cationic calcium and a carbonate/water component that occupies a connected channel network, resulting in a system that is inherently electrostatically instable [65]. The developed model exhibits a range of Ca coordination environments that encompasses those of the crystalline calcium carbonate polymorphs. Carbonate ions and water molecules are mainly coordinated by hydrogen bonding. It is assumed that the channel network of ACC is stabilised by various additives.

The transformation of ACC into calcite via the intermediate formation of vaterite is described in Ref. [67]. Calcium carbonate precipitation by mixing aqueous $1\,M$ Na_2CO_3 and $1\,M$ $Ca(OH)_2$ solutions was investigated in situ by time resolved ED-XRD. A two-stage crystallisation pathway was found between 7.5°C and 40°C. The first stage is the dehydration of ACC, resulting in the formation of nanocrystalline vaterite. The second stage, which is approximately 10 times slower, is the conversion of vaterite into calcite by means of dissolution and recrystallisation. The reaction rate of the second step is controlled by the surface area of calcite.

Ikaite ($CaCO_3 \cdot 6\,H_2O$): Metastable phase of calcium carbonate with monoclinic crystal system. It was found in nature for the first time at the bottom of the Ika Fjord in Greenland [68].

Monohydro calcite ($CaCO_3 \cdot H_2O$): Metastable calcium carbonate polymorph, trigonal crystal system. It has been reported that MHC was formed during the carbonation of nanolime suspension under higher RH [69].

Basic calcium carbonate ($2\,CaCO_3 \cdot Ca(OH)_2 \cdot n\,H_2O$): It crystallises in the orthorhombic system and decomposes in the presence of water to calcite and aragonite in the temperature range between 40 and 80°C. $Ca(OH)_2$ is also formed [70a].

At 25°C, the solubility of $CaCO_3$ polymorphs increases from calcite through aragonite, vaterite and ikaite to ACC. This means that all other polymorphs are metastable with respect to calcite. Their formation requires the presence of solutions with high super-saturations and/or the presence of inorganic/organic additives. In principle, conversion into calcite should take place over time. But it is a well-known fact that metastable polymorphs are able to exist over longer periods of time. Recently, it was shown that anhydrous crystalline polymorphs exhibit different nanomechanical properties [70b]. ACC also plays a major role in biological processes.

3.2.2 Hydraulic Lime

- Lime consisting mainly of calcium hydroxide, calcium silicates and calcium aluminates
- Reacts and hardens with water to form calcium silicates and/or calcium aluminate hydrates
- Reaction with atmospheric carbon dioxide is part of the hardening process

Hydraulic lime can be subdivided into the groups (in accordance with DIN EN 459-1:2015):

- Natural hydraulic lime, which is obtained through the calcination of limestone containing clay minerals or silica (including chalk) and subsequent slaking with water. Natural hydraulic lime is hydrated with sufficient water to convert calcium oxide into calcium hydroxide while the silicates or aluminates that are present remain unaffected.
- Hydraulic lime, which is obtained by mixing lime with materials such as cement, slag sands, limestone flour or other suitable materials.
- Formulated lime, which consists of air lime and/or natural hydraulic lime and contains other hydraulic and/or pozzolanic materials.
  Components can be:
  o Cements
  o Natural pozzolans
  o Other pozzolanic materials (e.g., calcinated clays such as metakaoline, silica fume)
  o Limestone powder
  o Slag sand
  o Additional additives
    ∗ Organic additives (e.g., plasticisers, flow improvers, air pore forming agents, hardening retarders, hardening accelerators, agents for hydrophobic properties)

* Mineral additives (e.g., additives such as Fe- or Sn-compounds used for the complexation of chromium)

Both air and hydraulic lime are used in various ways for the conservation of historic monuments, for example, as mortar, render or injection grouts. Their final physico-chemical properties depend on many factors. The characteristics of the raw materials used for their production, burning and slaking conditions as well as the binder-aggregate ratio or the water content are only few examples. Moreover, these materials have also experienced a 'renaissance' in modern constructions, especially on account of their environmentally friendly behaviour and the generation of a healthy indoor climate. Apart from typical lime slurries, repair mortars, injections grouts and fillers, the following materials are often used:

Lime wash (or whitewash): Thin calcium hydroxide slurry used to produce, protective layers on stone or renders. The thickness should be below 5 mm.

Dispersed lime hydrate: Suspension of air lime in water characterised by special properties due to the dispersion of the solids in water using high shear rates.

Lime varnish: Systems based on calcium hydroxide used to obtain transparent coatings. May be modified with lime-resistant pigments and sometimes contains organic additives.

3.3 Dissolved Inorganic Components Used for Consolidation

3.3.1 Lime Water

Lime water is the common name for a saturated solution of calcium hydroxide in water. The solubility of calcium hydroxide in water is low (approximately 1.65 g $Ca(OH)_2$ per 1 L water at 20°C). The solubility of $Ca(OH)_2$ decreases with an increasing temperature. Consequently, only small amounts of $Ca(OH)_2$, but high volumes of water are brought into treated substrates. Like all calcium hydroxide-based systems, strengthening is achieved by a reaction with atmospheric carbon dioxide.

Lime water is referred as a 'traditional' agent for restoring and consolidating degraded calcareous material in some literature. However, lime water is missing from the well-known list of consolidants drawn up by Friedrich Rathgen in 1898. For example, in the Czech technical literature, the method was added to the list of consolidants in 1953 [71] without any reference. Tracing the use of lime water in literature is complicated by the fact that the same term is also frequently used for lime wash [72].

Peterson [73] described lime water applications for restoring wall paintings, that is, plasters that have not disintegrated much since they are generally found indoors. References are not presented, but may probably draw on some older publications for example [74, 75].

A paper by Nicholas J. T. Quayle, from the Cliveden Conservation Workshop Ltd. [76] refers to a negative citation from previous literature [77, 78] as well as his own experience of limestone consolidation. He admits that he experienced positive effects with some 'soft' limestone (Bath, Doulting, Chilmark), but these were very inconsistent and unpredictable. Some parts of the surface were not consolidated at all, and this was typical for 'harder' limestone (Portland). Quayle's final argument against lime water technology concerns the mechanical removal of loosened particles from the surface during lime water application, which was also reported by Ashurst et al. [79] and later by Drdácký et al. [21, 80].

The systematic work on limestone was initiated by a reported application during the restoration of sculptures at Wells Cathedral (Somerset, England, UK) by Professor R. Baker. His restoration procedure, based on Peterson's recommendation of 40 applications of lime water wetting, was preceded by lime putty poultices and concluded with a sacrificial lime wash paint made of lime, sand and limestone dust, which greatly affects the final appearance of the restored monument. The paint is intended to be lost over time, but it temporarily protects the original surface and repairs from weathering.

Clifford A. Price [36] reproduced Baker's approach in the laboratory. To visualise new deposits of calcium hydroxide or carbonate in the limestone structure, he used radioactively-marked calcium ^{45}Ca [81]. He proved that 'some lime' deposits in the degraded limestone and the traces extended down to a depth of 26 mm. However, more than half of the lime remained in the 2 mm surface layer. Price also tested the mechanical effect (strengthening) on the treated material using abrasive blasting with quartz sand. No increase in material strength was detected. In another test, crushed limestone mixtures were treated. After 6 months, an increase of 0.38% and 0.40% in mass for two different types of limestone was found, but no consolidation effect. It was concluded that heavily disintegrated lime mortar cannot be consolidated with lime water through a few dozen wetting cycles. The work by Price that is described above is the most reliable and most serious study in the field of the penetration effect of lime water on various types of limestone.

The effect of lime water applied in situ to restore wall paintings on lime mortar rendering has been well investigated by I. Brajer and N. Kalsbeek [82]. These researchers systematically tested lime water treatments with respect to the application procedure, the number of cycles (20–70 cycles), dosage and maturation. They concluded that continuous 'wet' applications lead to a consolidation effect, unlike applications with 'drying' breaks, which do not consolidate the wall painting. No objective measurement of the mechanical characteristics was carried out during this study. Of course, the problem of fixing a thin layer of paint that has become detached from the substrate or pigment in the lime wash paint is totally different from that of mortar consolidation (which affects a bulky material). In this case, lime water may play an important role in the restoration practice. A laboratory study focusing on the consolidation treatment of a weak lime mortar using specimens that simulate degraded lime render [35] showed that lime water has to be applied repeatedly in multiple applications (sometimes more than one hundred times) for it to become significantly effective. If the render had not become too degraded, Tavares et al. [83] reported a sufficient improvement of its strength following applications of lime water. Extensive reviews of the lime water method have been carried out by Fiedler [84], Woolfit [85], Hansen et al. [86], Doehne and Price [87] as well as Siegesmund and Snethlage [17]. The recent study by Matteini et al. (2011) [88]

pointed out the very low calcium hydroxide content in lime water, which limits the efficiency of the product, and concluded that too many applications are required to obtain a significant consolidating result. This entails the risk of introducing an excessive amount of water into the system.

3.3.2 Barium Hydroxide

Barium hydroxide, unlike $Ca(OH)_2$, is highly soluble in water. At 25°C, solutions containing up to $40\,g/L$ $Ba(OH)_2$ can be prepared by dissolving compounds such as $Ba(OH)_2$ or $Ba(OH)_2 \cdot 8\,H_2O$. Consolidation can be achieved by carbonation in a manner similar to that of calcium hydroxide.

$$Ba(OH)_2 + CO_2 \longrightarrow BaCO_3 + H_2O$$

In many cases, however, other reactions dominate. $Ba(OH)_2$ forms insoluble $BaSO_4$ (barite) with sulphate ions. This is a fast reaction that takes place almost spontaneously on account of the low solubility of $BaSO_4$. This is often the reason why only low penetration depths are achieved, too. Mere traces of sulphate ions are sufficient to block pores and prevent deep migration.

Barium hydroxide solutions are able to convert gypsum into $BaSO_4$ and $Ca(OH)_2$, which will be present in either a dissolved form or as a solid (depending on the water to $Ba(OH)_2$ ratio and the gypsum content). It is a combined dissolution and precipitation process. In the first step, gypsum dissolves in water. Because $BaSO_4$ has a much lower solubility than gypsum, the sulphate ions immediately form barite precipitates.

$$CaSO_4 \cdot 2\,H_2O + Ba(OH)_2 \longrightarrow BaSO_4 + Ca(OH)_2 + 2\,H_2O$$

Under ideal conditions, for example when stirring fine gypsum crystals in a barium hydroxide solution, it is possible to convert gypsum completely into $BaSO_4$. Dissolution and precipitation take place locally and at different times. But such conditions do not prevail during the treatment of gypsum zones, for example in stone or in crusts on marble or limestone monuments. The great risk is that gypsum will only be converted on the surface. The $BaSO_4$ layer that forms then prevents a deeper penetration of the $Ba(OH)_2$ solution. It has been reported that barium hydroxide treatments are not appropriate for substrates that are rich in magnesium carbonates (e.g., plaster made with dolomitic lime) due to the formation of highly soluble magnesium compounds [86]. The application of barium hydroxide solutions has in many cases led to the formation of grey-white precipitates on the surface of treated objects.

$Ba(OH)_2$ is also used in the so-called 'Ferroni method' or 'Florentiner method', which are aimed at converting gypsum into $CaCO_3$ and $BaSO_4$. To achieve this, zones containing gypsum are initially treated with $(NH_4)_2CO_3$ solutions, leading to the formation of $CaCO_3$, according to the equation:

$$CaSO_4 \cdot 2\,H_2O + (NH_4)_2CO_3 \longrightarrow CaCO_3 + (NH_4)_2SO_4 + 2H_2O$$

The resulting ammonium sulphate $(NH_4)_2SO_4$ is converted into $BaSO_4$ through treatment with $Ba(OH)_2$ solutions in a second step.

$$(NH_4)_2SO_4 + Ba(OH)_2 \longrightarrow BaSO_4 + 2\,NH_3 + 2\,H_2O$$

Unreacted $Ba(OH)_2$ forms $BaCO_3$ through a reaction with atmospheric CO_2; ammonia theoretically evaporates completely. The successful application of this combined treatment in the conservation of wall paintings has been described several times, especially by Italian restorers. More information about the Ferroni method can be found, for example, in the recent book by Baglioni et al. [89] and Ref. [90].

The problems, however, are the same as for the treatment with pure $Ba(OH)_2$ solutions; impermeable layers can form that prevent the complete conversion of gypsum. In addition, it is difficult to remove ammonia completely. Over time, unremoved ammonia will turn into nitrate and new salt problems may arise.

Special attention has to be paid to the preparation of the barium hydroxide solutions. $Ba(OH)_2 \cdot 8\,H_2O$ or $Ba(OH)_2$ has to be dissolved with CO_2-free, deionised water in an inert atmosphere (preferably N_2 atmosphere). Due to the low solubility of $BaCO_3$ and its fast formation, contact with air alone is sufficient to produce insoluble barium carbonate precipitates. This results in suspensions of fine $BaCO_3$ particles, which have a much lower penetration capacity than pure barium hydroxide solutions.

Apart from this, all soluble barium compounds are highly toxic and special care is needed during their handling.

3.4 Consolidants Based on Silicates

3.4.1 Introduction

Silicates and sandstone are an important group of natural building materials and a main group of artificial stones. In cementitious systems, they are formed by the reaction of pozzolans with lime or cement through the hydration of hydraulic lime. In every case, the resulting silicates are crystalline, which means that they are characterised by well-defined mineralogical phases. The consolidation of silicates and sandstone can be achieved by:

- Fluorosilicates
- Alkali silicate solutions
- Colloidal silica
- Silicic acid esters (SAEs)

Apart from SAEs (see Chapter 6), typical properties and applications will be considered in the following in more detail. However, some general remarks are essential beforehand.

All of the components form colloidal silica gels, which act as a binder. These are inhomogeneous compounds consisting of condensed molecules of silicic acid. The gels are amorphous, meaning that a crystalline lattice does not form. During formation, they are able to incorporate many molecules of water without any chemical binding. In time, and depending on the environmental conditions, this incorporated water will be released. This process is called 'syneresis', which means a drying of the materials due to the release of water, accompanied by shrinkage. All materials that produce silica gels are affected by this process.

Crystalline silicates can only form when cementitious materials, meaning suspensions of reactive calcium-silicates, are used, excluding hydrothermal processes [17].

3.4.2 Fluorosilicates

Fluorosilicates are one of the oldest synthetic conservation mate-rials. They are the salts of hexafluorosilicic acid (H_2SiF_6). $MgSiF_6$ and $ZnSiF_6$ are primarily used on their own or in mixtures. They are applied as aqueous solutions in concentrations up to 2–4 wt%. The solutions are highly acidic ($pH < 2$), due to the formation of dissolved HF in contact with water.

The consolidation effect is based on the formation of silica gel (SiO_2) through a reaction with calcium hydroxide as follows:

$$2\,Ca(OH)_2 + MgSiF_6 \longrightarrow 2\,CaF_2 + MgSiF_2 + SiO_2(gel) + 2\,H_2O$$

Similar reactions take place following contact with calcium carbonate:

$$2\,CaCO_3 + MgSiF_6 \longrightarrow 2\,CaF_2 + MgSiF_2 + SiO_2(gel) + 2\,CO_2$$

A treatment with fluorosilicate solutions results in dense layers with a thickness of only a few millimetres. There is a great danger of the formation of crusts on weaker subsurfaces.

Treating sandstone with fluorosilicates may harm the stone due to the aggressive nature of the acidic solutions [17].

3.4.3 Soluble Alkali Metal Silicates

The use of materials leading to the formation of amorphous silicate gels has a long history. Alkali silicate solutions were regarded as suitable components, particularly at the beginning of the 20th century. Silicates are typical insoluble in water and mineral acids, but do dissolve in strong alkaline conditions. For example, melting sand with alkali carbonates results in the formation of water-soluble alkali-silicates, also called 'water glass'.

Water glass solutions are colourless aqueous systems with a mostly high viscosity. They are characterised by the weight ratio of silica to alkali, expressed as $SiO_2 : M_2O$. The $SiO_2 : Na_2O$ ratio of commercial sodium silicate solutions varies between 1.6 and 3.3. This corresponds with SiO_2 contents of between 25 and 32 wt% as well as Na_2O contents of between 9 and 16 wt%. $SiO_2 : K_2O$ ratios between 1.8 and 2.5 are typical for potassium silicates, resulting in SiO_2 contents of 20–27 wt% and K_2O amounts of 8–15 wt%. Both sodium and potassium silicate solutions are strongly alkaline (pH $\approx$ 11–13). All silicate solutions can be diluted with water in all ratios. The pH value, however, remains high. In commercial products the density is expressed in degrees Baumé (° Bé), which can be converted to conventional specific gravity by the following equation:

$$\text{Density} = 145/(145 - °\text{Bé}) \qquad (3.1)$$

The density depends on the content of dissolved SiO_2. The minimum viscosity of sodium silicate solutions is achieved at a $SiO_2 : Na_2O$ ratio of 2.0. The viscosity increases as this weight ratio becomes either more siliceous or more alkaline.

All water glass solutions are very complex systems that contain a wide variety of different silicate ions, including orthosilicate ions ($SiO_4{}^{4-}$), different linear and cyclic ions as well as crosslinked polysilicate ions. Their presence depends on the concentration, temperature and SiO_2-to-alkali ratio.

Neutralisation through the addition of acids or acidic salts causes gelation. Depending on the concentration and the composition of the neutralisation agent that is added, the mixing ratio and the temperature, silica sols or precipitated silicas may also form. However, this does not play a role in stone conservation treatments.

The application to consolidate stone is often described by the formal equation:

$$Na_2SiO_3 \cdot n\, H_2O + CO_2 \longrightarrow Na_2CO_3 + SiO_2 \cdot n\, H_2O$$

The rate of this reaction is slow and one cannot expect the carbon dioxide to penetrate pores that are completely filled with a water glass solution. It is likely that the formation of silica gels is mainly the result of drying, caused by the evaporation of water, or by soaking into the pore systems of the substrate. The penetration depth, even for diluted alkali silicate solutions, is often low (only a few millimetres). This is the result of interactions between dissolved monomeric and polymeric silicate ions and mineral surfaces, which lead to a fast gel formation and thus to a limited penetration. A dense skin often results, leading to a loss of adhesion and consequently to delamination. Another problem is that alkali silicate solutions lead to increased salt contents in the treated materials, especially on account of the formation of soluble carbonates. Today, potassium silicate solutions are mainly used as components for silicate paints. Another application is as a primer, especially for loose renders.

The application of lithium silicate solutions has been discussed over the past few years. Unlike sodium and potassium silicate, lithium silicate solutions are obtained by a simple dissolution of silica in LiOH solutions. Commercial products contain around 2.5 wt% Li_2O and 20.5 wt% SiO_2, corresponding to a weight ratio of 8.2. The viscosity is low (20 cp) compared to sodium and potassium silicate solutions. Thorn [91] describes the favourable application of lithium silicate solutions for the consolidation of wet stone and plaster, even under conditions in which silicic acid esters cannot be applied. One main advantage is the low solubility of Li_2CO_3 compared to Na_2CO_3 and K_2CO_3. Lithium silicate is also recommended for the preparation of injection grouts. Reports summarising complex conservations or large-scale applications, however, are unknown.

3.4.4 Colloidal Silica

Colloidal silicas are dispersions of amorphous, non-porous silicic acid particles in water. They are spherical and are not crosslinked, or only to a small extent. In contrast to water glass solutions, colloidal silica is characterised by a low alkali content. Typical particle sizes are in the range between 3 and 150 nm, depending on the type. The SiO_2 concentrations vary between 15 wt% and 50 wt%. Silica sols are characterised by a low viscosity and have specific surface areas of between 30 and 1000 m^2/g. They have a clear to opal/white appearance, depending on the concentration of dispersed silica.

The dispersions are obtained from sodium silicate solutions by means of ion exchange. Cationic exchange resins (loaded with H_3O^+) are brought into contact with sodium silicate solutions. The sodium ions are bound to the cation exchange resin, while protons are released. Apart from forming spherical silica nanoparticles, this process is accompanied by a significant reduction of the alkali content.

The stability of the dispersion results from electrostatic repulsion forces, whereby it is important to distinguish between anionic and cationic stabilised systems. Anionic stabilisation is achieved by the addition of hydroxide ions, for example NaOH, or as a result of ammonium hydroxide. Cationic stabilisation takes place when sodium aluminate is used.

Anionic stabilised dispersions are characterised by pH values between 8 and 10.5, whereas cationic systems have a neutral or slightly acidic character. Silica sols can also be stabilised by adding silanes.

If the pH value remains unchanged, anionic dispersions are stable against gelling and settling. Destabilisation occurs when the sols come into contact with salts and salt solutions (NaCl, $CaCl_2$, KCl, etc.) [92]. Weak gels are formed.

Unlike other systems that form silica gel, the use of colloidal silica dispersions is of only minor interest for stone and mortar conservation. Problems may occur during penetration into stone structures due to the formation of gel caused by the compensation of the electrostatic forces (breaking up of the dispersion), interactions with salts as well as purely water penetration into deeper zones. As result, only thin densified layers may be obtained.

Silica sols are put to successful use in the treatment of cementitious surfaces as well as porous renders and to prepare special repair mortars and silicate-based paints.

Another way to prepare silica sols is by the hydrolysis of silicic acid esters, in particular the hydrolysis of tetraethyl orthosilicate (TEOS). If this reaction takes place under defined conditions, it produces silica sols in ethanol with particle sizes between 10 and 100 nm. The favourable application of this kind of sol (with a SiO_2 content between 4 and 8 wt%) to consolidate sandstone has been described in Ref. [93]. Penetration depths up to 1.2 cm have been achieved in Saxon sandstone (Cotta sandstone). Nothing, however, has been reported about its long-term behaviour and successful treatment of various objects.

3.5 Other Consolidants

3.5.1 Ammonium Phosphates

Ammonium phosphates (diammonium hydrogen phosphate or ammonium dihydrogen phosphate) were proposed to consolidate calcareous substrates and were tested on limestone [88, 94–96] and mortars. The strengthening effect was significant (an almost 30% increase in tensile strength). The procedure was relatively fast (two days). Only non-hazardous aqueous solutions were used and a consolidation depth of more than 2 cm could be achieved. The freezing-thawing cycle test of the treated limestone samples showed that the stability was better than that of non-consolidated samples [96]. The efficiency of the ammonium phosphates was also tested on marble specimens [97].

The novel potential consolidant aluminium phosphate ($AlPO_4$) was recently investigated for the treatment of marble. A combination of diammonium hydrogen phosphate and aluminium phosphate was employed. The aluminium phosphate was used because of the greater similarity of the lattice parameters to calcite than hydroxyapatite formed by using $CaCl_2$ and ammonium phosphate solutions. The lattice mismatch with calcite is only 1% for aluminium phosphate compared to 5% for hydroxyapatite. In fact, treatment with aluminium phosphate significantly improved the dynamic elastic modulus of weathered marble specimens [98].

3.5.2 Ammonium Oxalate

Calcium oxalate layers can also be deposited on limestone surfaces analogous to ammonium phosphate solutions [78, 79]. This can be done in at least two ways. Ammonium oxalate can react with the carbonate stone or with render to produce a layer of calcium oxalate. Liu et al. [99] have analysed a 1000-year-old protective film on ancient inscriptions and then successfully prepared a similar coating through biomimetic synthesis.

Treatment with oxalic acid or their salts can bind calcium ions as stable calcium oxalate. Patinas are formed that act as a protective water repellent. Next, the surface of the stone is coated with a chondroitin sulphate (CS) film, followed by oxalate precipitation (CS induces the crystallisation). This procedure uses relatively inexpensive chemicals and, because the reaction occurs under mild conditions, it may also be useful to protect weathered lime mortars. However, to the best of our knowledge, no in situ test were performed using renders as substrates.

3.5.3 Tartaric Acid

Tartaric acid (HOOC-CHOH-CHOH-COOH) has an excellent solubility in water (at 20°C > 1000 g/L) and in different alcohols. Tartaric acid reacts with calcium carbonate at low pH values to form calcium tartrate tetrahydrate, which has a low solubility. It can grow on carbonate rock and the resulting layers strengthen the treated zones and protect the stone against acid attacks [100]. Due to the presence of hydroxyl groups in calcium tartrate tetrahydrate, an enhanced surface bonding of tetraethoxysilane-based consolidants as well as alkylalkoxysilanes has been found.

3.5.4 Biomineralisation

Another approach to consolidate stone and especially carbonates is the use of biological processes leading to the formation of calcite, something that has been discussed for many years. It is a well-known fact that certain bacteria have a substantial influence on inorganic systems. They are able to release metabolites and contribute cells (cell fractions, etc.), which can then act as heterogeneous nuclei for the crystallisation of inorganic minerals such as oxalates, silicates, carbonates, phosphates or sulphates. The formation of calcium carbonate is of particular interest for stone conservation. The most appropriate bacteria for this are heterotrophic aerobic or microaerophilic bacteria. These release ammonia, which dissolves in water and produces NH_4^+ and OH^- ions. Alkaline conditions are generated, under which atmospheric and metabolically-formed CO_2 is converted into carbonate/hydrogen carbonate ions. Calcium carbonate crystallisation occurs in the presence of calcium ions when the solubility product is exceeded. Again, the nature of the resulting calcium carbonate modification depends on specific conditions such as temperature, absolute supersaturation and the presence of other inorganic or organic substances.

Most bacteria are able to induce carbonate precipitation [101]. There are generally two ways in which bacteria can be used to deposit calcium carbonate. One of them involves the application of suspensions containing calcinogenic bacteria and nutrients (including calcium salt) onto the object [63], for example, the use of the Bacillus cereus to precipitate calcium carbonate, consolidating highly porous stones and filling microcracks in dense limestones [102]. The second approach seeks to activate the native microbial communities present within calcareous stones [103, 104]. A typical culture medium leading to enhanced bacterial growth is described in the following: 10 g/L Bacto Casitone (hydrolysed casein for enhanced bacterial growth), 10 g/L $Ca(CH_3COO)_2 \cdot 4 H_2O$ (calcium acetate), 2 g/L $K_2CO_3 \cdot 0.5 H_2O$, and 10 mM phosphate buffer in distilled water. Rodriguez-Navarro et al. [104] described the successful application of this solution for the consolidation of several buildings without any colour changes on the surface. The resulting degree of consolidation of calcarenite stone is the same as that achieved by using ethyl silicates.

Biomineralisation is an attractive option in conservation science with some undeniable achievements. Nevertheless, costs would appear to be the major factor preventing its wider use [105]. Biodeposition can also be used for artificial building materials such as cement-based mortars [106]. The treatment of such materials should be regarded as a coating since the penetration of bacteria into the pores is limited. Furthermore, it has to be mentioned that the nutrients needed may cause new salt damage and act as source of unwanted microbiological growth (fungi, algae).

3.6 Nanomaterials

A comprehensive review of multi-functional nanomaterials for timber in construction are discussed in Ref. [58b] beside nanomaterials in general can be found in Refs. [40, 107]. Apart from nanolime, the application of $Mg(OH)_2$, $Ba(OH)_2$ and $Sr(OH)_2$ nanoparticles is discussed together with the use of colloidal silica. Additional aspects concern the use of alkoxide nanoparticles, hydroxyapatite or silicon-based hybrid polymer nanocomposites and self-cleaning coatings.

The conclusion reached in Ref. [107] is that the use of nanotechnology offers possibilities for conservation treatments with enhanced material properties and novel functionalities. However, it also points out that it is essential to study the effectiveness, compatibility and durability of the new nanomaterials. This is imperative so as to prevent inadequate treatments.

3.7 Conclusion

The most important inorganic materials for the consolidation of stone, mortar and plaster are systems based on air lime or hydraulic lime. They are mainly applied as mortars or injection grouts. Their strengthening effect is based on the formation of calcium carbonate or calcium carbonate in combination with cementitious phases. The carbonation of calcium hydroxide through a reaction with atmospheric carbon dioxide is a very complex reaction that comprises a number of physical and chemical steps. Although calcite is the thermodynamically stable $CaCO_3$ modification, the formation of different metastable calcium carbonate polymorphs is possible. The carbonation of calcium hydroxide is associated with the release of water, but the presence of additional 'free' water is of significant importance for carbonation. Carbonation does not take place in dry environments.

Although the use of lime water is often proposed as a traditional method, it cannot be recommended for all substrates. It entails several disadvantages that could potentially be risky for the treated objects, especially with a very weak plaster/render or objects contaminated with salts. The solubility of calcium hydroxide in water is around 1.6 g/L at 20°C, resulting in the formation of a maximum of 2.2 g/L $CaCO_3$. This means that mostly water is brought into treated zones. Consolidation systems based on dissolved inorganic substances such as sodium silicate, fluorosilicate or barium hydroxide solutions may be of importance for special conservation tasks, but their use is often associated with several disadvantages. Their application is therefore very restricted, in particular with regard to sodium silicate and fluorosilicate.

Systems leading to surface modifications such as ammonium phosphate, ammonium oxalate or tartaric acid solutions are still in their development stage. One main problem with their application is the compatibility with historic materials.

Systems based on colloidal silica are promising, but the improvement to mechanical properties is generally low. Amorphous silica gels with low strengthening effects are formed.

It can be concluded that if silicic acid esters are excluded on account of their intermediate position between inorganic and organic consolidants, no liquid inorganic consolidants with a high penetration capacity into structurally degraded materials are available. One alternative is the use of dispersions of extremely fine calcium hydroxide particles ('nanolime'), which will be discussed in the next chapters in more detail.

Chapter 4

Fundamentals of Nanolime

Gerald Ziegenbalg, Claudia Dietze, and Gabriele Ziegenbalg

IBZ-Salzchemie GmbH & Co. KG, Schwarze Kiefern 4, 09633 Halsbrücke, Germany
gerald.ziegenbalg@ibz-freiberg.de; claudia.dietze@ibz-freiberg.de;
gabi.ziegenbalg@ibz-freiberg.de

4.1 Introduction

One of the main challenges in the conservation of the building environment is the correct choice of materials that are compatible with the original substrate as well as previous interventions. In both Europe and many other parts of the world, this primarily calls for materials that are compatible with natural and artificial stone. Materials and conservation techniques are needed to treat mainly inorganic materials, especially silicates, carbonates or sulphates. Ideally, the same components that were used in the original material will be formed during conservation. However, as already outlined in Chapter 3, the number of possible inorganic conservation materials is limited. The conservation of calcareous materials in particular is often a challenging task. Silicic acid esters, the standard material for structural consolidation, often cannot be used because the silica gels that form do not bind to calcareous surfaces.

Nanomaterials in Architecture and Art Conservation
Gerald Ziegenbalg, Miloš Drdácký, Claudia Dietze, and Dirk Schuch
Copyright © 2018 Pan Stanford Publishing Pte. Ltd.
ISBN 978-981-4800-26-6 (Hardback), 978-0-429-42875-3 (eBook)
www.panstanford.com

In principle, carbonates can easily be obtained by mixing solutions containing calcium and carbonate ions. The disadvantage of this method is that the crystallisation reaction occurs as soon as the two separate solutions come into contact. Often, extremely fine particles are formed. The major disadvantage, however, is that the resulting solution will contain dissolved salts such as NaCl, KNO_3, etc., depending on the composition of the solutions used.

$$CaCl_2 + Na_2CO_3 \rightarrow CaCO_3 + 2\,Na^+ + 2\,Cl^-$$
$$Ca(NO_3)_2 + K_2CO_3 \rightarrow CaCO_3 + 2\,K^+ + 2\,NO_3^-$$

One alternative is to use calcium hydroxide, which forms calcium carbonate through a reaction with atmospheric carbon dioxide. Unfortunately, calcium hydroxide has a low solubility ($\sim$1.65 g/L at 20°C) and suspensions have to be used to introduce sufficient binder into the materials to be consolidated. The consolidation of structurally deteriorated areas as well as powdery surfaces, however, requires materials that can penetrate fine fissures, cracks and pores. Nevertheless, dispersions of the finest calcium hydroxide particles in different alcohols ('nanolime') are an alternative to conventional conservation materials.

The first use of calcium hydroxide nanoparticles for the conservation of cultural heritage preservation were described by Ambrosi et al. [108] and Salvadori and Dei [109] in 2001. The particles, which were synthesised by precipitation in aqueous solutions, were used to stabilise wall paintings. The first commercially available nanolime product was CaLoSiL®, which was brought on the market in 2010. Today, many different CaLoSiL® products are available. These will be discussed in more detail below. A second nanolime product is available under the name Nanorestore Plus®; it contains 5 g/L or 10 g/L $Ca(OH)_2$ dispersed in ethanol or *iso*-propanol. The manufacturer is CSGI – Consorzio per lo Sviluppo dei Sistemi a Grande Interfase, via della Lastruccia 3, 50019, Sesto Fiorentino, Italy.

There are a number of ways to synthesise nanolime dispersions, such as:

- The precipitation of $Ca(OH)_2$ at room or elevated temperatures through a reaction with aqueous $CaCl_2$, $Ca(NO_3)_2$ and NaOH solutions

- The pecipitation of $Ca(OH)_2$ through a reaction of calcium salts with NaOH in organic solvents (often in ethylene or propylene glycol) or microemulsions [110, 111]
- The reaction of calcium hydride or carbide with water [112]
- A solvothermal reaction of metallic calcium with ethanol followed by the decomposition of resulting ethanolate with water
- Ion exchange [113]
- The reaction of elemental calcium with water followed by dispersion of the resulting calcium hydroxide particles in different alcohols
- Plasma deposition [114]

The first two methods are characterised by a problem that is also typical for $CaCO_3$ precipitation; the final solution contains dissolved salt, from which the calcium hydroxide particles have to be separated. One possibility is filtration and repeated washing with water, though this proves difficult due to the small particle size and the high viscosity of many organic solvents.

Calcium carbide decomposes rapidly into $Ca(OH)_2$ and C_2H_2. The resulting particles coagulate immediately. The same problems arise when CaH_2 is used. The solvothermal method, where the reactions take place in an autoclave system at a high temperature and pressure, is promising, but expensive and time-consuming.

The calcium hydroxide nanosols (CaLoSiL®), which will be discussed in the following section in detail, are obtained in a proprietary manner that uses complex reactions between water, metallic calcium and ethanol, whereby it is important that an exactly defined ratio is guaranteed between the different components.

Over the past few years, several papers have been published that deal with the application of nanolime dispersions. Many of these will be included in the following discussions.

4.2 Physico-Chemical Properties of Calcium Hydroxide Nanosols

Calcium hydroxide nanosols (CaLoSiL®) are white to opal, alcoholic dispersions with concentrations between 5 g/L and 50 g/L (Fig. 4.1).

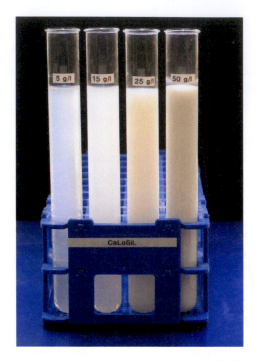

Figure 4.1 Colloidal dispersions of nanolime with 5–50 g/L in ethanol (CaLoSiL®).

Ethanol and mixtures of *iso*-propanol or *n*-propanol with ethanol serve as solvents. The standard products are summarised in Table 4.1.

The letters E, IP and NP stand for ethanol, *iso*-propanol (2-propanol) and n-propanol (1-propanol), respectively. The number indicates the concentration in g/L, for example, CaLoSiL® E25 means 25 g/L $Ca(OH)_2$ dispersed in ethanol. Whereas all CaLoSiL® E

Table 4.1   Standard CaLoSiL® products

Product	Solvent	$Ca(OH)_2$ conc.
CaLoSiL® E	Ethanol	5–50 g/L
CaLoSiL® IP	*Iso*-propanol/Ethanol	5–25 g/L
CaLoSiL® NP	*N*-propanol/Ethanol	5–25 g/L
CaLoSiL® grey	Ethanol	5–25 g/L
CaLoSiL® paste like	Ethanol	100 g/L
CaLoSiL® micro	Ethanol	120 g/L

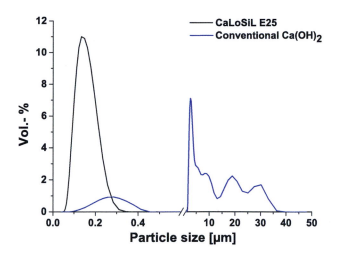

Figure 4.2 Particle size of CaLoSiL® E25 and conventional white lime hydrate in ethanol measured by laser scattering.

products contain only ethanol as a solvent, the IP and NP products are based on mixtures of the alcohols mentioned above and ethanol. This is a result of the specific synthesis conditions. Depending on the final concentration, the amount of ethanol in the NP- and IP-based products is not higher than 50 vol%. CaLoSiL® grey is a material that is characterised by a slightly grey colour. All of the products exhibit a maximum water content of 2 vol%. CaLoSiL® paste like and CaLoSiL® micro are also available. The first consists of slightly larger calcium hydroxide nanoparticles in ethanol in concentrations up to 100 g/L. CaLoSiL® micro contains 120 g/L $Ca(OH)_2$. The particle sizes are between 1–3 µm. Both CaLoSiL® paste like and CaLoSiL® micro are used to fill fractures, modify the other CaLoSiL® products or prepare special mortars and injection grouts.

Figure 4.2 shows the typical particle size distribution for a nanosol containing 25 g/L $Ca(OH)_2$ in ethanol. The particle sizes range between 50 nm and 250 nm. It is obvious that the mean particle size of CaLoSiL® E25 is 2 orders of magnitude smaller than that of conventional lime hydrate. Furthermore, the distribution range of the nanolime is considerably narrower than that of conventional lime hydrate. The particles dispersed in *iso*-propanol/ethanol or in *n*-propanol/ethanol mixtures are characterised by similar particle size distributions. They are plate-like hexagonal crystals, which is

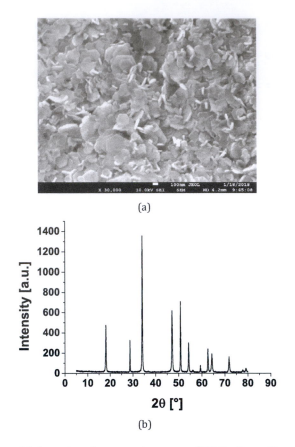

(a)

(b)

Figure 4.3 (a) Scanning electron microscopy (SEM) of nanolime. (b) XRD pattern of nanolime.

typical for $Ca(OH)_2$ (Fig. 4.3a). Although the particles are extremely fine, they are crystalline and exhibit the typical XRD patterns of portlandite (Fig. 4.3b).

Additional characteristics are summarised by Rodriguez-Navarro [66]. The length of the particles (measured along [110]) ranges from 35 to 235 nm (average ± std. deviation 134 ± 57 nm). Their thickness (measured along [001]) is between 15 and 40 nm (average ± std. deviation: 25 ± 8 nm). In some cases, aggregates of a few (3–12) particles with a size of 300–600 nm have been observed

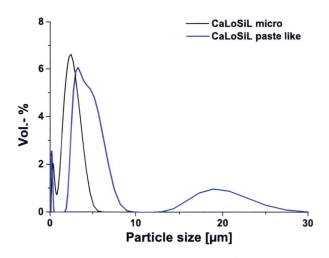

Figure 4.4 Particle size distribution for CaLoSiL® paste like and micro measured by laser scattering.

by TEM. The size of the nanoparticles determined by laser scattering is 34–400 nm (mode 138 nm). The BET surface was determined as $31.4 \pm 0.5 \, \text{m}^2/\text{g}$ [66].

Typical particle size distributions of CaLoSiL® micro and CaLoSiL® paste-like, measured by laser scattering, are shown in Fig. 4.4. The morphology of CaLoSiL® micro is shown in Fig. 4.5.

Figure 4.5 Morphology of CaLoSiL® micro.

Table 4.2 Thermodynamic properties of water, ethanol, *n*-propanol, *iso*-propanol at atmospheric pressure and 25°C [115–117]

	Water	Ethanol	*N*-propanol	*Iso*-propanol
Molecular weight [g/mol]	18.02	46.07	60.10	60.10
Density [g/cm^3]	0.99	0.79	0.80	0.78
Dynamic viscosity [mPa·s]	0.89	1.20	1.95	2.07
Boiling temperature [°C]	100.0	78.4	97.2	82.0
Vapour pressure at 20°C [hPa] (www.dgvu.de)	23.0	58.0	20.3	42.6
Surface tension [mN/m] (www.surface-tension.de)	72.8	22.1	23.7	23.0
Dielectric constant	80.37	25.00	20.81	18.62

Whereas $Ca(OH)_2$ has a low solubility in water, it is almost insoluble in ethanol, *n*-propanol or *iso*-propanol and in mixtures of alcohols. Typical thermodynamic properties of the three different alcohols compared to water are shown in Table 4.2. As the data indicates, there are significant differences between the alcohols and water, though also between the alcohols themselves.

The low boiling points of the alcohols and corresponding high vapour pressures result in rapid evaporation (Fig. 4.6).

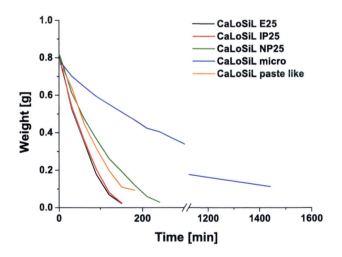

Figure 4.6 Evaporation behaviour of different CaLoSiL® products.

Table 4.3 Density and viscosity of nanolime dispersions depending on the concentration of $Ca(OH)_2$ and the solvent

CaLoSiL®	Density [g/cm^3]	Dynamic viscosity [mPa·s]
E50	0.83	1.8
E25	0.81	1.6
E5	0.79	1.3
IP25	0.81	1.9
IP5	0.79	2.3
NP25	0.82	1.9
NP5	0.80	2.1

Compared to water, ethanol as well as *n-* and *iso*-propanol are characterised by a lower density but a higher viscosity. The presence of calcium hydroxide nanoparticles results in an increased viscosity and density. The absolute increase, however, remains low (Tables 4.2 and 4.3). In this context, it should be noted that the products CaLoSiL® IP and NP are production-related mixtures with ethanol in the respective ratios. Accordingly, the dynamic viscosity of IP25 (solvent IP : E 1 : 1 vol%) is lower than that of IP5 (solvent IP : E 10 : 1 vol%).

The wetting behaviour of surfaces is primarily determined by the surface tension of the corresponding solution, which is caused by cohesive forces between the molecules. Molecules have no neighbouring molecules above them on their surface and consequently exert stronger attractive forces on their closest neighbours. The surface tension is defined as the property of a liquid's surface to resist external forces (see Fig. 4.7).

Liquids with a low surface tension generally tend to wet solid surfaces better. Due to the use of alcohols as solvents, the nanolime dispersions have a lower surface tension than water, resulting in an excellent wetting behaviour. When nanolime dispersions are placed on glass plates, for example, complete wetting takes place. The evaporation of the alcohol leads to the formation of a dense layer of calcium hydroxide. The wetting behaviour is directly related to the capillarity of porous or fractured materials. This will be discussed in more detail in Section 4.2.2.

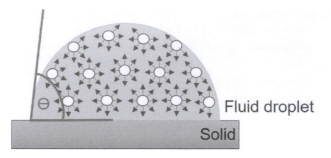

Figure 4.7 Schematic illustration of the surface tension with the characteristic contact angle Θ between the tangent of the fluid surface to the solid surface.

When particles are dispersed in a solvent, interactions take place between the solid surface and the solvent. These interactions determine whether stable dispersions are formed or sedimentation takes place. In a polar solvent, the surface spontaneously acquires a surface electrical charge. The main mechanisms are the ionisation of surface groups, adsorption of charged species (ions, ionic surfactants) and specific ion dissolution from crystal lattices.

Two layers are formed on the particle surface:

- One that is strongly connected to the surface
- A second one that is characterised by weak electrostatic forces

In this double layer, the first layer is called the Stern layer and describes strong interactive forces between a negatively charged particle surface and positively charged cations. In the outer zone, the cations are only bonded weakly. The forces of attraction between the negatively charged particle surface and the cations in the solution decrease with an increasing distance from the surface (Fig. 4.8).

The zeta potential (ζ-potential) characterises the charge that develops at the interface between the solid surface and the surrounding solution. It is a measure of the electrical charge of particles dispersed in a solvent. The zeta potential is not equivalent to the electric surface potential in a double layer or to the Stern potential but describes the double-layer properties of a colloidal dispersion. Typically, the higher the zeta potential, the more stable the colloidal systems will be. When the zeta potential equals zero, the dispersed solids will coagulate. The zeta potential is expressed

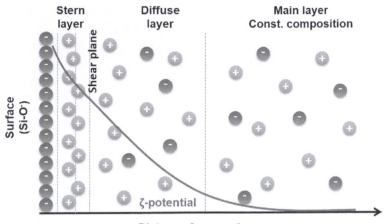

Figure 4.8 Schematic illustration of the electrochemical double-layer.

in milli-volts (mV). It is determined by the surface charge of the particles, any layer adsorbed at the interface, and the nature and composition of the surrounding solvent.

Similar ionisation effects are possible in solvents with moderate dielectric constants (>10). Charging mechanisms such as those in water occur. Stable dispersions are formed when strong repulsion forces are active. In contrast, attraction forces between the particles result in the formation of agglomerates and floccules.

Calcium hydroxide dispersions in all three alcohols are characterised by a positive zeta potential as shown in Table 4.4. Thus, the stability is expressed by the electrostatic repulsion forces of calcium hydroxide particles surrounded by positively charged calcium ions combined with steric effects. As long as repulsion forces remain intact, the sols do not settle.

Table 4.4 Zeta potential of different CaLoSiL® products (Zetasizer instrument; Malvern)

CaLoSiL®	Zeta potential [mV]
E5	57.8
IP5	52.4
NP5	22.8

The stability of nanolime dispersions is greatly influenced by the presence of water as well as any salts. Adding up to 5% of water does not influence the short-term stability of the dispersion to a great extent. The addition of an excess of water, however, results in coagulation and flocculation of the calcium hydroxide particles (see Fig. 4.9a). Calcium hydroxide gels as well as large calcium

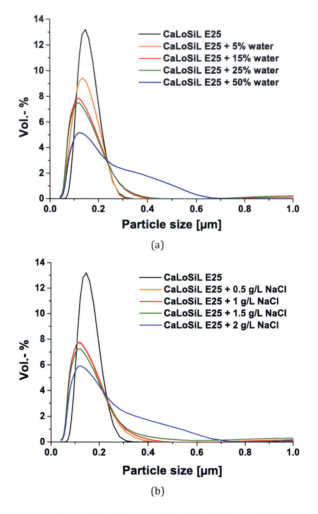

(a)

(b)

Figure 4.9 Particle size distribution after the addition of (a) water and (b) NaCl measured by laser scattering.

hydroxide agglomerates are formed, depending on the amount of water that is added. The addition of water enhances the solubility of $Ca(OH)_2$, resulting in an decrease of the zeta potential and thus to a compensation of the repulsion forces. Although NaCl is not soluble in ethanol, its effect on the particle size is very similar to that of adding water (Fig. 4.9b).

4.2.1 Sedimentation Stability

An excellent way to characterise the stability of nanolime dispersions is to measure the sedimentation rate using an ultracentrifuge. To this end, the nanolime dispersion is placed in a transparent cuvette and accelerated up to $2300\,g$, resulting in enhanced sedimentation. A special analyser measures the transmission of light through the sample during acceleration. Zones of well-mixed dispersions scatter and adsorb the light so that transmission is low. In contrast, any clarification allows more light to reach the detectors and the transmission increases. This zone grows with time and acceleration, indicating the sedimentation process. The sedimentation characteristics during centrifugation are an indication of the stability of nanolime dispersions. The principle is illustrated in Fig. 4.10.

Typical results are shown in Fig. 4.11, where the sedimentation rate is plotted against the rotation speed, expressed as a relative centrifugal force (RCF). The more gentle the slope of the graphs, the lower the sedimentation rate and the more stable the dispersions are. Ideally, a line parallel to the x axis is obtained, which would characterise a perfectly stable system. It is quite obvious that the stability of nanolime dispersions is determined by both the solvent used and the concentration of the calcium hydroxide nanoparticles. In general, the higher the concentration of the dispersed calcium hydroxide particles, the lower the stability of the nanolime dispersion. Dispersions in *iso*-propanol/ethanol mixtures are characterised by the highest stability with the same content of $25\,g/L\ Ca(OH)_2$.

The measurements correlate well with the determination of the particle size distribution over time. Measurements with calcium

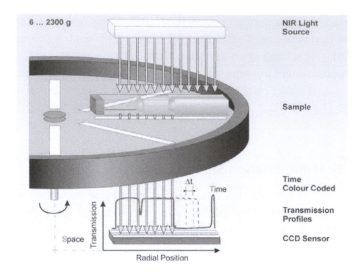

Figure 4.10 Measuring principle of the ultracentrifuge Lumifuge®, provided by LUM GmbH.

hydroxide dispersions of 15 g/L in ethanol showed complete stability against sedimentation over a period of three years (Fig. 4.12).

Comprehensive investigations of more concentrated dispersions in ethanol (25 g/L, 50 g/L) showed that the sols are stable for at least six months. When sedimentation does occur, the particles can be re-dispersed mechanically, for example by shaking. The sedimented solids are characterised by the same particle size as the dispersed solids. In contact with water, however, rapid sedimentation and coagulation takes place, as seen in Fig. 4.9.

4.2.2 Penetration Behaviour of Calcium Hydroxide Nanosols

Structural consolidation as well as the strengthening of powdery or delaminating surfaces requires the consolidation agent to penetrate the substrate. Whereas the flow of a liquid into large openings is mainly gravity driven, transportation into small pores, fractures and fissures is caused by capillarity. This is the ability of a narrow flow channel (tubes, fissures, pores) to draw a liquid upwards against the force of gravity.

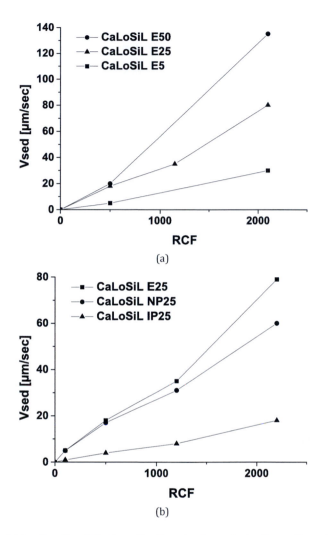

(a)

(b)

Figure 4.11 Results of the investigations to characterise the sedimentation stability of nanolime dispersions using an ultracentrifuge. Sedimentation time is plotted against the relative centrifugal force (RCF). (a) CaLoSiL® E at different concentrations (5, 25, 50 g/L), (b) CaLoSiL® with different alcohols (E–ethanol, NP–*N*-propanol, IP–*Iso*-propanol).

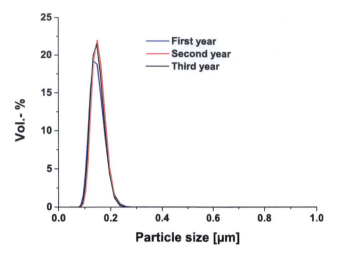

Figure 4.12 Development of the particle size distribution of CaLoSiL® E15 over three years.

The capillarity of porous media results from two opposing forces (see Fig. 4.13):

- The adhesion of the liquid to solid surfaces
- The cohesive surface tension of liquids that acts to reduce the liquid-gas interfacial area

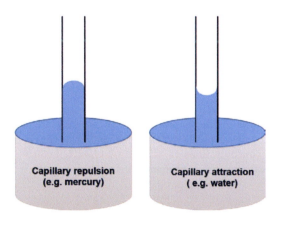

Figure 4.13 Capillary action.

Capillarity requires the solution to wet the surface. A concave meniscus forms (see Fig. 4.13 right). The narrower the channels or pores, the higher the solution rises. If wetting does not take place, a convex meniscus forms (see Fig. 4.13 left) and there is no capillary rise. On the contrary, a capillary fall is observed. The capillary rise can be described by the following equation:

$$h = \frac{2 \cdot \gamma \cdot \cos \theta}{r \cdot g \cdot \rho} \tag{4.1}$$

h–height of a liquid column; γ–surface tension; θ–contact angle; ρ–density of the liquid; r–radius of the capillary; g–local acceleration due to gravity.

If it is assumed that glass is completely wettable by water ($\theta = 0$) and the surface tension of water is 0.072 N/m, the capillary rise depending on the diameter of the capillary can be expressed by the equation:

$$h \approx \frac{1.47 \cdot 10^{-5} \text{m}^2}{r} \tag{4.2}$$

It is clear that the capillary rise is higher in thin capillaries than in larger ones (see Fig. 4.14a). If ethanol ($\rho_{\text{ethanol}} = 0.785$) is used as a solvent and a surface tension of 0.0221 N/m is assumed, the following approximation can be used to determine the capillary rise:

$$h \approx \frac{0.58 \cdot 10^{-5} \text{m}^2}{r} \tag{4.3}$$

The capillary rise of ethanol is lower than that of water (see Fig. 4.14).

(a)   (b)

Figure 4.14 Glass capillaries with different inner diameters. (a) CaLoSiL® E25; (b) water coloured in blue.

The capillary uptake of water (and of any other solution) is a time-dependent process, that can be described by the following equation:

$$x(t) = B \cdot \sqrt{t} \tag{4.4}$$

x–capillary rise [m]; B–water penetration coefficients, expressed in [m/s$^{1/2}$] describes the rate of capillary rise. The value of B depends on the capillary pressure and the flow resistance.

Similarly, the amount of adsorbed water *($m_w(t)$ in [kg/m^2])* is described by a $\sqrt{time}$ law.

$$m_w(t) = A \cdot \sqrt{t} \tag{4.5}$$

The coefficient A, which is expressed in kg/(m$^2 \cdot$ h$^{1/2}$) describes the rate of water uptake until the maximum height is reached.

It is generally assumed that suspensions are able to penetrate porous materials when the particle size of the solids is between three and five times smaller than the pore diameter. The maximum particle size of nanolime in CaLoSiL® is 250 nm. This means that pores with a clear diameter of 1.25 µm or more should be penetrable. It has to be remembered, however, that this is a theoretical value. Interactions between the particles and surrounding surfaces as well as with any moisture or salts that are present may result in coagulation processes as well as the blockage of large pores or fractures.

Nanolime generally displays a good penetration behaviour with depths of several centimetres. In Ref. [118], the penetration of CaLoSiL® E25 into different porous UK limestones has been investigated in detail. The main results are summarised in Table 4.5.

In most cases, the penetration depth is more than sufficient to achieve an appropriate consolidation. The main problem is the possible reverse-migration of the nanoparticles during the evaporation process (see also Chapter 5).

4.3 Carbonation of Nanolime Dispersions

4.3.1 Introduction

Following evaporation of the alcohol, nanolime dispersions form Ca(OH)$_2$ layers on treated surfaces or within pore structures. In

Table 4.5 Rates and distance of capillary advection in three different limestones (from Ref. [119])

Limestone type	Rate of advection of CaLoSiL® E25 [cm/min]	Rate of advection of water [cm/min]	Water absorption [wt%]
Weldon	0.45	1.4	8.5
Ketton	0.1	0.4	6.7
Clipsham	0.02	0.08	7.7

Limestone type	Porosity [vol%]	Density [g/cm³]	Depth of penetration of CaLoSiL® E25 [cm]
Weldon	18.4	2.0	5.5
Ketton	15.4	2.14	4.5
Clipsham	17.5	2.1	4.2

time, these are converted into $CaCO_3$ through a reaction with atmospheric carbon dioxide. Although this process seems to be simple and is often described by the overall equation

$$Ca(OH)_2 + CO_2 \rightarrow CaCO_3 + H_2O$$

it is a multi-step reaction that is influenced by a number of variables. What's more, it has to be remembered that $CaCO_3$ exists in different modifications and variable morphologies (see Chapter 3).

The carbonation of aqueous $Ca(OH)_2$ slurries has been the subject of numerous investigations [120], especially in order to find favourable conditions for the industrial synthesis of different calcium carbonate modifications.

The following reaction sequence during the carbonation of lime mortars is probable:

- Diffusion of $CO_{2(g)}$ through the porous structure of the mortar.
- Dissolution of $Ca(OH)_2$ in the pore water.

$$Ca(OH)_2 \rightarrow Ca^{2+}_{(aq.)} + 2\,OH^-_{(aq.)}$$

- Reaction of gaseous CO_2 with the alkaline pore solution.

$$OH^- + CO_2 \rightarrow HCO_3^-$$
$$HCO_3^- + OH^- \rightarrow H_2O + CO_3^{2-}$$

- Precipitation of $CaCO_3$.

$$Ca^{2+} + CO_3^{2-} \rightarrow CaCO_3$$

The limiting step in the formation of calcium carbonate is the absorption of gaseous CO_2 in water. $CaCO_3$ may dissolve in the presence of an excess of $CO_{2(g)}$.

$$CaCO_{3(s)} + CO_2 + H_2O \rightarrow Ca^{2+} + 2\,HCO_3^-$$

Similar reactions take place during the carbonation of the calcium hydroxide formed after the evaporation of alcohol from nanolime dispersions. This means that the carbonation of $Ca(OH)_2$ is based on complex dissolution-crystallisation phenomena. The process requires the presence of water. Carbonation is an exothermic reaction that is accompanied by an increase in mass and volume and, as the overall equation shows, by the release of water.

The kinetics, and in particular the structure of the resulting calcium carbonates depend on several factors such as the pH value, degree of supersaturation, temperature, hydrodynamic conditions as well as the type and amount of impurities. In conventional crystallisation processes (for example mixing of $CaCl_2$ with Na_2CO_3 solutions), the presence of Mg^{2+}, Ni^{2+}, Co^{2+}, Zn^{2+} or Cu^{2+} ions promotes the formation of aragonite, whereas the presence of Mn^{2+}, Cd^{2+}, Sr^{2+}, Pb^{2+} or Ba^{2+} induces calcite formation [121]. The presence of alcohols further complicates the already very complex system because it has a significant influence on both the crystallisation kinetics and the resulting $CaCO_3$ polymorphs. Due to the importance of the carbonation step, the reactions of $Ca(OH)_2$ in water as well as in alcohols will be discussed in the following chapters in detail.

4.3.2 Carbonation of Lime in Aqueous Dispersions

Calcium hydroxide (portlandite) has a limited solubility in water. As Fig. 4.15 shows, the solubility decreases with an increasing

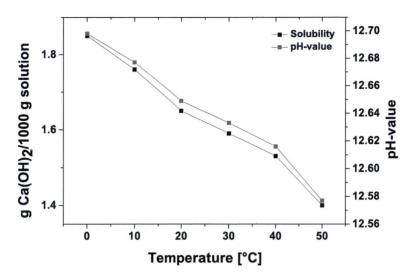

Figure 4.15   Solubility of $Ca(OH)_2$ in water and the respective pH value of the solution.

temperature, accompanied by a slight decrease in the pH value. As expected, the pH value depends on the amount of dissolved calcium hydroxide.

The solubility of solids depends not only on temperature and the overall solution composition, but in the case of nanoparticles on their size too. The solubility increases exponentially with the reduction of the particle size. This is a special state of metastability. The solutions are supersaturated with respect to coarse particles. The solutions will in time precipitate coarse particles. In the case of $Ca(OH)_2$, the solubility product can be described by the following equation:

$$K_{(Ca(OH)_2, s)} = (5.214 - 0.124 \cdot T) \exp^{(0.0313/d)} \qquad (4.6)$$

Here d is the particle diameter in μm and T the absolute temperature. It is mentioned in Ref. [122] that the concentrations obtained from experimental data were always lower than the theoretical values. The solubility was also found to depend on the amount of dispersed $Ca(OH)_2$. Nevertheless, the equation allows a rough estimate of the influence of particle size on solubility. Particles with a diameter of 150 nm (such as CaLoSiL®) are more than 20% more soluble than particles with a diameter of 2 μm.

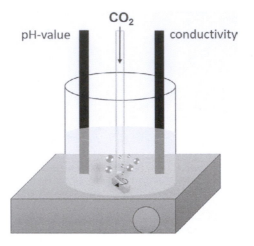

Figure 4.16 Schematic representation of the setup for the carbonation experiments.

To characterise the carbonation process of aqueous lime slurries in more detail, isothermal stirring experiments were carried out at room temperature. Carbon dioxide was introduced into a tank reactor at a constant volume flow (25 g/L $Ca(OH)_2$, 54 L/h CO_2). During this process, the pH value, conductivity and temperature were continuously recorded over 2 h. Figure 4.16 shows the experimental set-up.

The results are reproduced in Fig. 4.17. The development of conductivity can be divided up into the following stages:

(1) A sharp decrease within the first few minutes and the formation of a 'conductivity valley'. It is assumed that the surface of the calcium hydroxide particles are covered with calcium carbonate crystals within the first minutes and that these prevent further dissolution processes. Ref. [123] reports the formation of amorphous calcium carbonate layers (ACC). Conversion of ACC by means of dissolution and recyrstallisation into calcite (indicated by the increase in conductivity) causes cleaning of the surface and allows further calcite crystallisation.

(2) In the second stage, following the dissolution of the ACC layer, carbonation takes place at a nearly constant rate. As

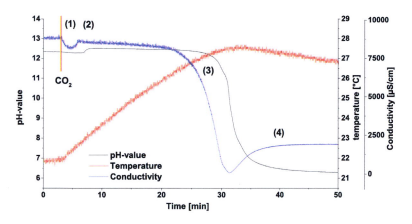

Figure 4.17 Conductivity, pH value and temperature change during the carbonation of $25\,g/L$ $Ca(OH)_2$ in an aqueous solution. The orange line indicates the beginning of the carbon dioxide flow, which was set to 54 L/h.

the conductivity indicates, the concentrations in the solution remain nearly constant, a sign of supersaturation.

(3) The sharp drop in conductivity indicates the end of the reaction.

(4) Further bubbling of CO_2 through the solution leads to the formation of hydrogen carbonate according to:

$$2\,H_2O + CO_2 \to HCO_3^- + H_3O^+$$

The overall curve for the pH value corresponds to the conductivity, whereas the slight increase in conductivity at the end of the reaction characterises the $CaCO_3/CO_2/H_2O$ equilibrium. The exothermic character of the reaction is indicated by an increase in temperature from 20 to 28°C. Figure 4.18 shows the development of conductivity depending on the volume flow of carbon dioxide. In all of these cases, calcite was found as final calcium carbonate phase.

It is generally accepted that the crystallisation of $CaCO_3$ follows Ostwald's step rule, meaning that metastable phases often form that are then gradually converted into the thermodynamically stable modification calcite [124]. This process depends on the temperature, available additives and the humidity. At room temperature and in the absence of additives, the carbonation of calcium hydroxide begins with the formation of amorphous calcium carbonate, which converts into vaterite or calcite.

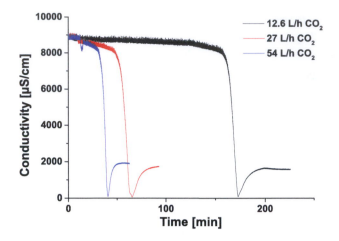

Figure 4.18 Change of conductivity during the carbonation of 25 g/L $Ca(OH)_2$ in an aqueous solution at a carbon dioxide volume flow of 54, 27 and 12.6 L/h.

High pCO_2 and alkaline conditions (pH 8.5–10.5) favour the formation of vaterite. Low pCO_2 and a pH value close to neutral or >11 favours the development of calcite crystals [124].

Strongly alkaline conditions prevail in a purely aqueous medium. ACC formation is the first step in the carbonation of $Ca(OH)_2$. It is formed as spherical or hemispherical nanoparticles which typically precipitate heterogeneously on $Ca(OH)_2$ plates [107]. The stable calcite phase forms by way of dissolution and subsequent crystallisation.

In the early stage, carbonation and evaporation of the water in the mortar overlap. Due to the slow diffusion rate of CO_2 in water, this stage is characterised by carbonation, which only takes place on the surface of the mortar.

The carbonation rate depends on many factors. Apart from the drying process and relative humidity conditions, the nature of the pore system also plays an important role. It determines the capillary transport regime and possible capillary condensation. Shrinkage will occur if the evaporation rate is too high due to a low RH (<60%), resulting in the formation of microcracks. The evaporation rate also determines the flux of water from the bottom towards the surface of the lime paste, which influences the amount of water remaining in the mortar structure and thus directly the carbonation process.

The importance of the presence of a liquid-like water adsorption layer for the carbonation of calcium hydroxide is summarised in Ref. [125]. To this end, porous calcium hydroxide particles were equilibrated with water vapour in defined relative pressure ranges at 20°C. It was discovered that liquid-like water adsorption layers catalyse the reaction between the calcium hydroxide particles and CO_2. The resulting solid phase enhances the evaporation of water so that the initial calcium hydroxide powders need to be equilibrated with water vapour at a relative partial pressure ≥ 0.7 to significantly promote the catalytic reaction. Surprisingly, it was discovered that the new solid phase is a non-protective layer of calcium carbonate, which is distributed entirely inside the initial porous particles. A minimum number of four layers of water adsorbed on the $Ca(OH)_2$ particles is needed. The necessity of the presence of adsorbed water to initiate the carbonation of $Ca(OH)_2$ was also confirmed by [126]. It was found that the gas-solid carbonation of portlandite is activated exclusively by initially adsorbed water molecules.

The following steps are assumed:

- Instantaneous activation by adsorbed water
- Autocatalytic step though the production of molecular water
- Passivation step though the formation of a protective carbonate layer around the core of reacting portlandite crystals

The dissolution and carbonation of portlandite single crystals are described in detail in Ref. [127]. It is concluded that the carbonation of portlandite, in both an aqueous medium and a high relative ambient humidity, takes place by means of a tight interface-coupled dissolution-precipitation process.

4.3.3 Carbonation of Alcoholic Calcium Hydroxide Dispersions

4.3.3.1 Solubility effects

The presence of alcohols changes the physico-chemical conditions for carbonation significantly because:

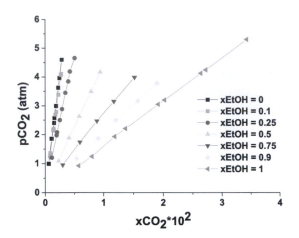

Figure 4.19 The solubility of CO_2 in ethanol-water mixtures at 298 K (data from Ref. [128]).

- The solubility of both calcium hydroxide and all calcium carbonate modifications decreases with an increasing alcohol content.
- In contrast to what has been mentioned above, the solubility of carbon dioxide in alcoholic solutions is much higher than in water.

Figure 4.19 shows the solubility of carbon dioxide in water and water-ethanol mixtures depending on the CO_2 pressure at 298 K (data from Ref. [128]). It is obvious that there is a linear relationship between the pressure and the mole fraction of the gas in the liquid phase. Although the CO_2 pressures are nowhere near the values that can be expected under atmospheric conditions, the data shows that ethanol is a much better solvent for carbon dioxide than water.

The solubility of $Ca(OH)_2$ in ethanol, *iso*-propanol and *n*-propanol is shown in Fig. 4.20 (from Ref. [123]). It is clearly visible that small amounts of the alcohols already result in a significant reduction of the $Ca(OH)_2$ solubility.

Possible interactions between alcohols and suspended calcium hydroxide particles are discussed in detail by Navarro et al. [129]. Conventional white lime hydrate was dispersed in ethanol as well as *iso*-propanol and then stored in closed bottles at temperatures

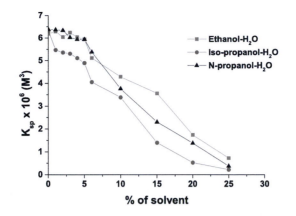

Figure 4.20 Ca(OH)$_2$ solubility at 293 K in mixed solvents (data from Ref. [123]).

between 40°C and 60°C for up to 60 days. CaLoSiL® E25 was used for comparative investigations. It could be shown that partial conversion of calcium hydroxide into calcium ethoxide (Ca(OC$_2$H$_5$)$_2$) takes place, especially at higher temperatures. The degree of conversion is low at around 3% and 7% at 40°C and 80°C, respectively. Similar reactions take place with *iso*-propanol, though to a much lesser extent (the degree of conversion is between 2% and 5%). Storing CaLoSiL® E25 for two weeks in a closed bottle at 60°C resulted in a degree of conversion into calcium ethoxide of approximately 8%.

The carbonation of CaLoSiL® E25 was almost complete within 48 h. A mixture of calcite with small amounts of vaterite had formed. In contrast, the carbonation of the samples containing calcium alkoxides was only 88% complete after two weeks. Vaterite was formed in this case. The adsorption behaviour of water and ethanol on calcite was investigated in Ref. [119]. In ethanol, well-ordered layers formed on the planar faces of calcite. Water exhibits a much weaker attachment. This means that ethanol adsorption is favoured over water.

4.3.3.2 Course of carbonation

The course of carbonation over time is greatly influenced by the presence of alcohols. Bubbling CO$_2$ through nanolime dispersions

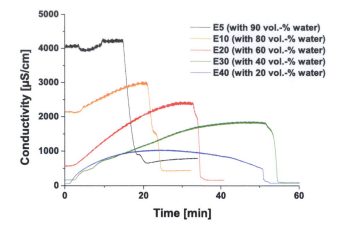

Figure 4.21 Course of conductivity during the carbonation of different nanolime slurries in water-ethanol mixtures (E40: 80 vol% CaLoSiL® E50 + 20 vol% H_2O; E5: 10 vol% CaLoSiL® E50 + 90 vol% H_2O, etc.).

(Fig. 4.21) results in a course of conductivity, that differs significantly from that obtained in aqueous systems (Fig. 4.18). The nanolime dispersions E5–E40 (5–40 g/L $Ca(OH)_2$) were prepared by diluting CaLoSiL® E50 (50 g/L $Ca(OH)_2$ in ethanol) with water, leading to different ethanol/water ratios and $Ca(OH)_2$ concentrations.

The 'conductivity valley' is only visible in systems that contain large amounts of water. As expected, the absolute value of the conductivity decreases with an increasing ethanol content due to the reduced $Ca(OH)_2$ solubility. The presence of only 20 vol% water (E10) already results in a significantly changed course of the conductivity. A continuous increase followed by a sharp decrease can be observed. The increase can be attributed to the increased solubility of carbon dioxide in ethanol-water mixtures. This means that a high degree of calcium carbonate supersaturation is likely. Thermogravimetric investigations of the reaction products indicated that the carbonation begins immediately after the CO_2 starts to flow through the nanolime-water-ethanol dispersions (Fig. 4.22). ACC forms, which is converted into crystalline polymorphs during complete carbonation. It is a well-known fact that ACC can

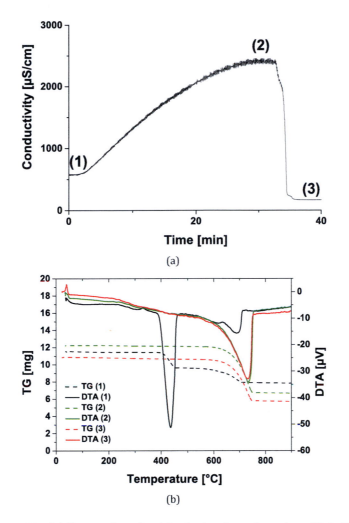

Figure 4.22 (a) Course of conductivity during the carbonation of CaLoSiL®
E20 (E20: 40 vol% CaLoSiL® E50 + 60 vol% H_2O). Number (1–3) indicates
the sampling points for the DTA/TG analysis. (b) DTA/TG graphs for the
respective samples (1–3).

be obtained when the precipitation processes are realised in
alcoholic solutions (see Ref. [130]). The exact composition of
the resulting ACC is unknown, especially if solvent molecules are
incorporated into its structure.

The complexity of the carbonation process in the presence of alcohols is discussed at great length in Refs. [69, 131]. CaLoSiL® IP was diluted to 2 g/L. The resulting solution was exposed to 33%, 54%, 75% and 90% relative humidity (RH), for 7, 14, 21, and 28 days without a CO_2 flow and no air circulation in climate chambers. Exposure to low humidity conditions (33% RH) resulted in slow carbonation processes. What's more, unreacted portlandite was present together with short prismatic crystals of vaterite after 28 days. At 54% RH, clusters of vaterite-calcite polymorphs (72 to 227 nm), which are associated with amorphous calcium carbonate (ACC), were found after 28 days. The crystallinity of the carbonates that formed increased significantly from 33% RH to 54% RH.

Exposure to 75% or 90% RH resulted in the formation of different species. Whereas $Ca(OH)_2$ (portlandite) had already disappeared within the first few days, calcite, aragonite, vaterite and monohydrocalcite (MHC) together with ACC were still found. The investigations at 90% RH in particular have shown that complex conversions take place between different calcium carbonate phases. The data in Tables 4.6 and 4.7 summarises the crystalline phases that were discovered, their relative content and their particle size distribution.

Table 4.6 Crystalline phases obtained during the contact of $Ca(OH)_2$ dispersions (CaLoSiL® IP25 diluted to 2 g/L) at 90% relative humidity (RH) [69]

Time [days]	Phase	Shape	Particle size [nm]	%
7	Calcite	Rhombohedral	40 ± 20	100
14	Calcite	Rhombohedral	91 ± 16	21
	Vaterite	Acicular	112 ± 38	17
	Aragonite	Interleaved sheets	608 ± 385	48
	MHC	Clay-like	1300 ± 200	14
21	Calcite	Rhombohedral	830 ± 180	72
	Aragonite	Rosette-like	$1.9 \pm 0.8\ \mu m$	19
	MHC	Clay-like	$4.9 \pm 1.5\ \mu m$	9
28	Calcite	Rhombohedral	563 ± 264	100

Table 4.7 Crystalline phases obtained during the contact of $Ca(OH)_2$ dispersions (CaLoSiL® IP25 diluted to 2 g/L) at 75% relative humidity (RH) [69]

Time [days]	Phase	Shape	Particle size [nm]	%
7	MHC	Clay-like	350 ± 100	100
14	Calcite	Rhombohedral	463 ± 170	10
	Vaterite	Acicular	110 ± 49	65
	Aragonite	Fibre-like	1755 ± 718	15
	MHC	Clay-like	750 ± 200	10
21	Calcite	Rhombohedral	616 ± 180	5
	MHC	Clay-like	900 ± 200	65
28	Calcite	Rhombohedral	400 ± 100	45
	Vaterite	Prismatic	179 ± 31	50
	Aragonite	Columnar	$2.2 \pm 0.5 \, \mu m$	5

The two papers [123, 128] confirm the dominant influence of humidity on the carbonation process and consequently the resulting calcium carbonate modifications. All of the investigations have indirectly shown that carbonation, and especially the conversion of the different calcium carbonate modifications, is based on coupled dissolution and precipitation processes. Ref. [69] states, 'Local fluctuations in the water/alcohol ratio significantly affect the precipitation/dissolution of anhydrous and hydrated polymorphs, as reflected in the particle size.' It should be mentioned that the crystallinity of the resulting calcium carbonates was significantly higher during exposure of the nanolime dispersion to 90% RH than to 75% RH.

The investigations summarised in Refs. [66, 132] clearly show the influence of alcohols on the crystallisation of $CaCO_3$ polymorphs.

When $CaCl_2$ and Na_2CO_3 solutions are mixed, for example, the dominance of calcite or aragonite can easily be controlled by the amount of alcohol added and the time of reaction [132]. In addition, mechanical aspects such as the shaking speed have a great influence on the morphology of the resulting calcium carbonate modifications. The transformation of ACC into calcite via the intermediate formation of vaterite is also described in Ref. [67].

It was found that after mixing aqueous $CaCl_2$ and Na_2CO_3 solutions in the temperature range between 7.5°C and 25°C, crystallisation takes place in the following two stages:

(1) The resulting particles of ACC dehydrate rapidly and crystallise to form individual particles of vaterite.
(2) Vaterite is transformed into calcite by means of a dissolution and recrystallisation mechanism with the reaction rate being controlled by the surface area of calcite.

The second stage of the reaction is approximately 10 times slower than the first.

A special aspect in the carbonation of nanolime dispersions is the possible presence of calcium alkoxides [66]. It has been found that the degree of conversion (with a constant time) decreases with increasing amounts of calcium alkoxides. Whereas the carbonation of calcium hydroxide always results in the formation of calcite, samples containing calcium alkoxide often convert into vaterite.

The kinetics of nanolime carbonation have been discussed in detail in Ref. [66]. To this end, CaLoSiL® E25 was deposited on a glass slide and subjected to drying at room temperature for 1 h. The samples were then placed in a plastic container at 18°C ± 2°C and 80% ± 5% RH. A small but continuous air flow promoted carbonation. The early stage of carbonation was characterised by the formation of amorphous calcium carbonate (ACC). Up to 13–24 wt% ACC formed during the first 2–4 h, which required the presence of humidity. Parallel tests in which the samples were stored under absolutely dry conditions did not result in any ACC formation within 2 months. In the second step, vaterite and calcite started to crystallise after 6 h together with traces of aragonite. The amount of vaterite increased rapidly over the first 24 h of carbonation. It then decreased while the calcite content increased. This was associated with a constant decrease in the calcium hydroxide content. In parallel, the aragonite content increased during the first day and decreased afterwards. After 21 days, no ACC remained in the sample. The amount of vaterite, also decreased drastically, whereas the mass of calcite increased accordingly. Overall, the carbonation of nanolime follows the sequence ACC – vaterite – aragonite – calcite and confirms Ostwald's step rule (from the less stable phase to

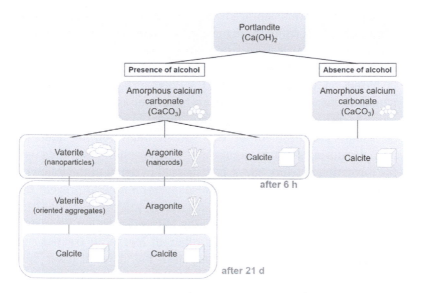

Figure 4.23 CaCO$_3$ crystallisation in the absence and presence of alcohol (adapted from Ref. [66]).

the most stable phase). It is assumed that the presence of alcohol favoured the formation of the metastable polymorphs vaterite and aragonite. The scheme shown in Fig. 4.23 was developed in Ref. [66].

Several investigations have confirmed that the metastable CaCO$_3$ phases are converted into other modifications by means of a coupled dissolution/precipitation process. This means that primary crystallised phases dissolve and the stable calcite modification or another metastable phase crystallises from the solution. This process is also greatly influenced by the composition of the solution in which conversion occurs.

4.4 Conclusion

Nanoparticles of calcium hydroxide can be stably dispersed in ethanol, *iso*-propanol or *n*-propanol. The resulting sols are characterised by particle sizes between 50 and 250 nm. Nanolime dispersions with concentrations between 5 and 50 g/L are commercially available. Their stability is based on electrostatic repulsion forces. Under normal storage conditions, CaLoSiL® dispersions are stable for at least 6 months in the absence of water. If water is added, gel formation and/or flocculation takes place.

The transport of the nanolime dispersions into porous structures is based on diffusion. Due to the low surface tension of the nanolime dispersions, the wetting behaviour is always excellent. The penetration depth of nanolime dispersions is, in most cases, sufficient. The problem of the reverse-migration of nanolime particles during solvent evaporation, which is often associated with the formation of white hazes, is discussed in detail in the following chapters. Recommendations are also given on how to prevent this phenomenon.

Evaporation of the alcohol results in the deposition of calcium hydroxide particles. These form adherent layers on treated substrates. Crystalline portlandite has been found in all cases after the evaporation of the alcohol. The following aspects have to be considered when discussing the carbonation of nanolime:

- Carbonation is always a complex process involving many physical and chemical steps. Important parameters include temperature, relative humidity, the structure of the substrate, the presence of impurities and the amount and structure of the resulting calcium hydroxide.
- It is a well-known fact that particles in the nanometer range have a higher solubility than normal, 'coarse' particles. This means that higher supersaturations are possible, as well as different reaction mechanisms, than during the carbonation of 'standard lime hydrate'.
- The solubility of carbon dioxide in ethanol and ethanol-water mixtures is much higher than in pure water.

All of these points taken together result in very complex reaction mechanisms and it comes as no surprise to learn that more or less all possible calcium carbonate polymorphs have been found in investigations carried out by several scientists. The following reaction sequence is probable:

- In the presence of high humidity, treatment with nanolime dispersions, following evaporation of the excess alcohol, results in alcoholic-water-solution films that form on the surface or inside the treated materials.

- The high solubility of the small nanolime particles and the high solubility of carbon dioxide produce conditions under which high supersaturations can be achieved. As a result, amorphous calcium carbonate (ACC) forms rapidly in the first step. Ikaite or MHC can also be expected to form under extreme conditions. ACC is converted into different calcium carbonate polymorphs depending on the reaction conditions. Again, the presence of water-alcohol films produces conditions of high supersaturation, encouraging the formation of metastable calcium carbonate polymorphs (vaterite, aragonite) on their own or in combination with stable calcite. The metastable polymorphs are kinetically stabilised by the presence of alcohol.
- The metastable calcium carbonate polymorphs are quickly converted into stable calcite under high humidity conditions. This process also takes place via dissolution-crystallisation steps, demonstrating the importance of high humidity. In some cases, small amounts of unreacted metastable calcium carbonate polymorphs may remain for long periods of time. This does not have a negative effect on the long-term stable consolidation of treated substrates because of their small amount (only a few percent).

It has to be remembered that the dissolution of the metastable calcium carbonate polymorphs in the adhering alcohol-water films may lead to the partial migration of the solution films back to the surface. This is not the case if the surrounding atmosphere has a high relative humidity.

Chapter 5

Laboratory Characterisation of the Action of Nanolime

Arnulf Dähne,[a] Christoph Herm,[b] and Gerald Ziegenbalg[c]

[a]*pons asini, Schloss 17, 04600 Altenburg, Germany*
[b]*Hochschule für Bildende Künste Dresden, Güntzstrasse 34, 01307 Dresden, Germany*
[c]*IBZ-Salzchemie GmbH & Co. KG, Schwarze Kiefern 4, 09633 Halsbrücke, Germany*
arnulf.daehne@pons-asini.de; herm@serv1.hfbk-dresden.de;
gerald.ziegenbalg@ibz-freiberg.de

5.1 Introduction

The characterisation of the action of stone or mortar consolidants by laboratory tests is always a challenging task that may result in conclusions that are difficult to reproduce under field conditions. The reasons for that are manifold:

- Laboratory tests are realised normally under ideal conditions, at constant temperature and humidity without negative effects such as intensive sunshine, rain or wind.
- Often 'synthetic' samples are used. That means, samples are applied which are composed by mixing of expected mortar compositions or by intact, not deteriorated stone.

Nanomaterials in Architecture and Art Conservation
Gerald Ziegenbalg, Miloš Drdácký, Claudia Dietze, and Dirk Schuch
Copyright © 2018 Pan Stanford Publishing Pte. Ltd.
ISBN 978-981-4800-26-6 (Hardback), 978-0-429-42875-3 (eBook)
www.panstanford.com

- In the laboratory it is always difficult to prepare samples reproducing damage characteristics and natural contaminations (e.g., salt contamination) completely. Often only partial reproduction of typical damage characteristics are possible.
- Due to the historic-cultural importance of many objects, often only small samples can be used for laboratory investigations. It is always difficult to assess if they are representative for the whole object or if they characterise only a special part.
- In the laboratory, careful treatments of smallest parts of an object are possible. During practical applications, however, often large areas are to be treated and also economic questions play an important role.

Notwithstanding the above, laboratory tests are of great importance in order to obtain an overview of the characteristics of conservation materials and their behaviour during application. Whenever possible, original samples should be used under conditions that are similar to which the object is exposed. Overall characteristics of the conservation materials can be determined by simple tests on representative components. The following chapters present principal possibilities of laboratory testing and ways to modify nanolime dispersions.

5.2 Favourable Laboratory Tests

The application of calcium hydroxide nanosols is possible by

- Injection
- Spraying
- Immersion, capillary suction, vacuum suction
- Brushing

The actually used method depends on the characteristics of the material to be treated. Main questions that can be in the focus of laboratory tests are:

- Achievable increase in the mechanical strength depending on the number of treatments
- Penetration depth

- Most favourable nanolime sol and its concentration
- White haze formation
- Most favourable application technique
- Required time until the final strength is achieved
- Possible combinations with other conservation materials

A first impression about the characteristics of conservation materials is obtained by determining the penetration behaviour into different materials. A simple possibility is to drop the consolidant on the surface of the material in question. This can be realised on real samples (stone, mortar, plaster) as well as on artificially prepared samples. Often, mortars with low binder contents are used to characterise the achievable increase in compressive and tensile strength. The penetration depth can be visualised by spraying of treated surfaces with an ethanolic 1 wt% phenolphthalein solution, which acts a pH sensitive indicator. The solution is colourless in the pH range between 0 and 8.7, at higher values a purple colour occurs. It is also possible to pretreat the material (for example mortar prism) with phenolphthalein solutions and to follow the capillary rise by the colour change caused by the migration of the calcium hydroxide nanoparticles (see Fig. 5.1).

Figure 5.1 Phenolphthalein indicating the diffusion of CaLoSiL® E25 into a mortar sample.

Figure 5.2   Sandwich samples. Photo: Strotmann & Partner.

The consolidating effect on powdery surfaces can be determined by peeling tests [20] before and after the treatment with the nanosols. One problem that often occurs in conservation is the stabilisation of fissures and delaminations. For that purpose, so-called 'sandwich' samples can be used. These consist of two pieces of intact stone (or mortar) and loose stone powder between them (Fig. 5.2).

The consolidant is injected directly into the loose material. The consolidation effect can be determined by measuring the adhesive strength or by drilling resistance measurements.

A favourable way to characterise the achievable connection between consolidated loose stone or mortar powder and surrounding

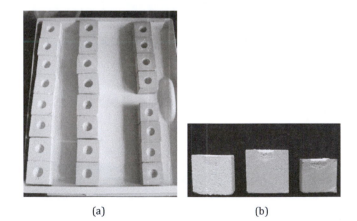

(a)   (b)

Figure 5.3   (a) Drill holes in intact stone. (b) Cross section of the samples after consolidation. Developed by Malgorzata Musiela (Restauro, Poland).

intact material is shown in Fig. 5.3. A hole is drilled into the material in question. The obtained drilling flour is given back into the hole, where it is saturated with the nanolime sol in the next step. This can be repeated until the required consolidation is achieved. Again, drilling resistance measurements are a favourable tool to characterise the obtained consolidation. Also optical methods, especially microscopy, are helpful to determine the success of the treatment.

The 'Karsten tube' is often used to measure the water absorption of porous structures. It can also be used for the determination of the volume of nanolime dispersions necessary to saturate a substrate.

As mentioned already, laboratory tests can give only first indications about the applicability of a conservation material, but they are absolutely necessary for the development of detailed conservation concepts. Laboratory tests are extremely helpful to find the suitable way of treatment. Additionally, they allow the modification of the materials, for example by addition of fillers or special admixtures. Other questions that should be investigated are possibilities of pre- or after-treatment with ethanol/water and the tendency of the consolidant to form a white haze on the surface of the treated material.

5.2.1 Modification of CaLoSiL® Nanosols Matching Specific Conservation Demands of Deteriorated Plaster and Stucco Work

Requirements for the conservation of plaster and stucco result from specific deterioration patterns of various characteristics and dimensions. Conservation demands comprise treatments ranging from structural strengthening of the porous structures to the removal of larger defects such as fissures, cracks, detachments and delaminations. Solvent-based calcium hydroxide dispersions open up the possibility of almost water-free and material appropriate conservation of mortar, for example if salt content or humidity-sensitive paint layers are present. Furthermore, the use of water-free lime-based grouts for the conservation of wall paintings, plaster and stucco is of great interest.

In preliminary treatment tests, however, standard calcium hydroxide nanosols of different concentrations proved not be suitable for the consolidation of lime mortar with high porosity. An increase of mechanical strength could not be stated and the surface showed a white haze due to accumulation of the consolidant. In order to achieve a penetration behaviour which is adequate to the pore structure of the substrate, the composition of the nanosol as well as its application procedure was to be adjusted. Generally, this can be achieved by dilution, additives, or after-treatment of the surface. For further purposes the nanosols were modified with additives, fillers, and aggregates.

5.2.2 Structural Strengthening of Mortar

For use as strengthener the consolidant generally has to meet several requirements, foremost sufficient increase of mechanical strength, homogeneous depth distribution, preservation of porosity and water absorption behaviour, and avoidance of colour change of the surface, for example, white haze. In order to achieve these objectives, various solvent compositions, the addition of thickeners, aggregates and lime dispersions with greater particle diameter were tested. The effect of the pure and modified calcium hydroxide nanosol strengthener was tested on very weak dolomitic lime mortar specimens as well as very weak gypsum mortar specimens, both resembling materials needing to be treated. The model materials used for the laboratory tests were:

- Dolomitic lime mortar prepared from DL 85-S (EN 459-1) and quartz sand (0.04–1 mm), binder : aggregate ratio = 1 : 32 mass/mass
- Gypsum mortar prepared from crushed used gypsum mortar and quartz sand (0.04–1 mm), final binder : aggregate = 1 : 24 mass/mass

The calcium hydroxide dispersions used were CaLoSiL® E25/IP25/NP25 each with 25 g/L as well as CaLoSiL® micro (particle size 1–3 μm, 120 g/L in ethanol). The nanosols were modified with 40% acetone, hydroxypropyl cellulose gel (0.5% mass in *n*-propanol), or CaLoSiL® micro (1% mass). The sols were applied by immersion or by pipette until saturation. The specimens cured at least seven days after application at 20°C/ 65% RH.

Figure 5.4 Distribution of calcium hydroxide nanoparticles in mortar samples indicated by the red colour of the phenolphthalein indicator (size $= 2$ cm $\times$ 2 cm), left: immediately after saturation with CaLoSiL® NP25, right: after 24 h.

The carbonation of the calcium hydroxide absorbed by the mortar was made visible by staining with phenolphthalein solution (0.1% m). For testing the mechanical strength three-point bending tests were performed on mortar prisms ($2 \times 2 \times 10$ cm^3) (average of three specimens). The water accessible porosity was determined by capillary water absorption under atmospheric pressure. More experimental details can be found in the literature [133].

After three times treatment with calcium hydroxide nanosol E25 the dolomitic lime mortar showed the accumulation of calcium hydroxide at the surface after evaporation (see Fig. 5.4). By comparison to the state immediately after impregnation it can be concluded that the accumulation is caused by back-migration of the nanoparticles during the evaporation of the solvent. The microscopic investigation showed the presence of consolidants in the pores, but also surface enrichment (see Fig. 5.5). In order to improve the distribution of the nanosol into the mortar and to avoid white haze on the surface the application method was modified in four directions:

(1) Use of diluted nanosols and a stepwise increase of concentration (5–25 g/L) during up to six applications
(2) Addition of acetone which already was reported in the literature with promising results [134]

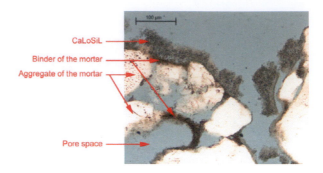

Figure 5.5 Microscopic image of a thin section (//Pol) of weak dolomitic lime mortar (laboratory specimen) after treatment with nanosol (CaLoSiL®), after complete curing. Photo: Thomas Köberle.

(3) Coating of the treated surfaces with ethanolic hydroxypropyl cellulose (HPC) gel in order to retard evaporation

(4) Combination of CaLoSiL® nanosol with coarser calcium hydroxide suspensions (CaLoSiL® micro), 'bimodal dispersion'

In each case, absorption of consolidant has reached 2–3 wt% (after drying) which equals an increase in binder content by up to 76 wt%. Dilution (1), combined with gel-coating (3) reduced the evaporation effectively and led to an even distribution (Fig. 5.6, left). The addition of micrometer-sized calcium hydroxide suspension

Figure 5.6 Distribution of calcium hydroxide nanoparticles in mortar samples after 24 h (phenolphthalein indicator) (size = 2 cm × 2 cm), left: modification of the solvent (CaLoSiL® NP12.5, 40% acetone) and after-treatment with HPC gel in ethanol (6-1), right: bimodal dispersion containing CaLoSiL® micro and CaLoSiL® E25 (7-1).

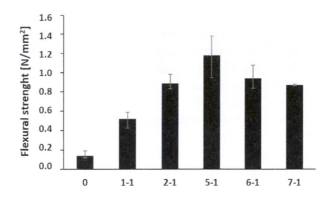

Figure 5.7 Three-point bending strength of weak dolomitic lime mortar (laboratory specimens, average of 3) before and after treatment with CaLoSiL® nanosols. 0 = untreated, Consolidants: 1-1 = E25 (3x), 2-1 = E(5–25 g/L, 6x), 5-1 = E (5–25 g/L) + 40% acetone (5x), 6-1 = NP (12.5 g/L) + HPC gel (6x), 7-1 = E25 + micro 100 + 1 w/w (3x).

(4) proved effective in preventing back-migration as well (Fig. 5.4, right).

The bending strength of the very weak dolomitic lime mortar $(0.14\,N/mm^2)$ was increased significantly by three treatments with CaLoSiL® E25 (increase of 271%, see Fig. 5.7, 1-1). The use of IP25 and NP25 resulted in a slighter increase of mechanical strength (increase of 142% and 250%, respectively). The modified application of the nanosols resulted in even higher mechanical strength of the mortar with almost similar increase of more than 500% compared to the untreated material: diluted nanosol (5–25 g/L) (2-1), and coating with HPC gel after application of nanosol diluted with acetone (6-1). The combination of stepwise increased concentration and acetone admixture show the best results (742% compared to the untreated material and about double increase of the strength of the pure E25 treatments, see Fig. 5.7, 5-1). Finally, the combination of CaLoSiL® nanosols with coarser calcium hydroxide suspensions (CaLoSiL® micro) resulted in significantly increased mechanical strength (186%–521%, see Fig. 5.7, 7-1). The water accessible porosity of the dolomitic lime mortar after treatment with these various calcium hydroxide nanosol formulations and application methods was only slightly reduced from 36–39 vol% (untreated) to 30–34 vol% (treated).

For the very weak gypsum mortar specimens, generally the same approach was made as for dolomitic lime mortar. In each case, absorption of consolidant has reached 2–3 wt% (after drying) which equals an increase in binder content by up to 70 wt%. Similar to the dolomitic lime mortar the bending strength of the gypsum mortar prisms could be increased by treatment with the calcium hydroxide nanosol strengtheners. The bending strength ($0.114\,N/mm^2$) was increased significantly by three treatments with CaLoSiL® E25 (increase of 339%). The use of IP25 and NP25 resulted in slighter increase of the mechanical strength (increase of 198% and 233%, resp.). The modified application of the nanosols resulted in even higher mechanical strength of the gypsum mortar (increasing concentration of diluted ethanolic nanosol: 418% increase of strength; nanosol diluted with 40% acetone: 549% increase). The combination of stepwise increased concentration and acetone admixture showed the best results (increase of strength of 680% compared to the untreated material).

5.2.3 Consolidation of Cracks and Gaps

A grout on the basis of CaLoSiL® nanosol already available on the market (CaLoXiL® injection grout) still contains a fraction of water. The aim was to develop suitable grouting masses using different products of the CaLoSiL® product line without any water and admixtures, as discussed below.

For the consolidation of thin layered delamination and fixation of fine cracks and fissures, unmodified nanodispersions are not suitable because of their low viscosity. The necessary retention of the applied nanosol in small spaces should be enhanced by modification in three directions:

(1) Modified solvent composition or addition of gelling agent (hydroxypropyl cellulose Klucel® G, 0.5% mass in ethanol + water [1+1 v/v])
(2) Use of filler with very low particle size: natural calcium carbonate (max. 30 μm), hollow glass microspheres (Scotchlite™ S22, 75 μm and Scotchlite™ K1, max. 120 μm)
(3) Dispersion with 'bimodal' particle size distribution using admixtures of CaLoSiL® micro

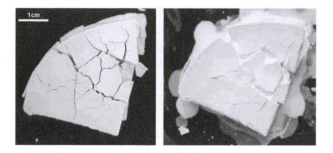

Figure 5.8 Artificially aged limestone sample before (left) and after (right) treatment with modified nanosol.

Another task was to find possibilities for the consolidation of delaminations and for fixation of cracks and fissures with a width above 1 mm. Here, pure nanosols are not suitable because of their low viscosity and insufficient strength. Preliminary tests showed that thickening of the nanosol with hydroxypropyl cellulose only did not lead to any improvement. Increased binder content has resulted in shrinking-cracks and decrease of strength. Promising results were found using CaLoSiL® micro mixed with CaLoSiL® E50 as binder. Furthermore, for increase of viscosity and prevention of shrinkage the use of a filler has proven necessary. More experimental details can be found in the literature [133].

Limestone samples: In order to simulate a heavily damaged substrate with very fine cracks and fissures, limestone samples were used for laboratory tests (see Fig. 5.8). Burning of a fine micritic limestone at 700°C and subsequent slaking using damp sand led to multilayered delaminations and severe cracks with a width up to about 1 mm. The consolidants were applied three times using a syringe. The specimens cured 20–30 h after application at 20°C/ 65% RH).

The consolidation was evaluated based on optical appearance and stability against manual brushing. The best results were achieved by the admixture of hydroxypropyl cellulose solution with water content. Although necessary, the water content can be reduced to a total of 2.5 vol%. Despite the good consolidation effect, only very fine defects were filled by this agent after up to three applications. For filling larger gaps and fissures of 0.5–1.0 mm width, a second test series was carried out using higher concentrated

CaLoSiL® (50 g/L) mixed with CaLoSiL® microdispersion. After a pre-treatment with nanosols modified as described above, the mixture of CaLoSiL® E50 and CaLoSiL® micro (50 : 50 v/v) with addition of hydroxypropyl cellulose solution in ethanol and water proved to be suitable for completing the consolidation. A higher content of nanosol in the mixture showed in turn worse results.

In order to simulate filling of fissures and delaminations the grout was applied between two brick cubes with a contact surface of c. 2×2 cm^2. The specimens cured 7 days after application at 20°C/ 65% RH. For comparative studies a commercial water-based lime-grout was used (PLM-A, c.t.s, Altavilla Vicentina/ Italy). The adhesive strength was tested using a universal mechanical testing machine (three specimens per mixture).

Mixtures with 1 : 1 v/v ratio of CaLoSiL® nanosol and CaLoSiL® micro gave the best results. Here, the ratio of nanoparticles and microparticles is c. 30 : 70 m/m. A filler mixture was calculated in order to give dense particle packing (chalk or marble powder $+$ S22 $+$ K1 $= 72\% + 16\% + 12\%$ m/m). The grout composed of $6 + 6$ mass fractions of CaLoSiL® E50 and CaLoSiL® micro and 1 mass fraction of filler added resulted in an adhesive strength of 0.046 N/mm^2 see Fig. 5.9 H-29). This value reached c. 50% of the

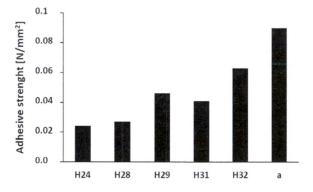

Figure 5.9 Adhesive strength (pull-off test, N/mm^2) of grout-bonded brick samples (maximum values of three specimens). Mixture test: CaLoSiL® micro $+$ CaLoSiL® E50 $+$ filler (m/m/m): H24 $= 4 + 2 + 1$, H28 $= 4 + 4 + 1$, H29 $= 6 + 6 + 1$, H31 $= 8 + 8 + 1$, H32 $= 4 + 4 + 1$ (marble powder), a $=$ commercial grout (PLM-A).

strength of a commercial grout (PLM-A, see Fig. 5.9, a) and thus is comparably high and suitable for many demands. Using marble powder instead of chalk for the filler mixture improved the adhesive strength to 0.063 N/mm^2 (i.e., c. 70% of the commercial grout, see Fig. 5.9, H32). The best performing mixtures also developed sufficient adhesive power to the mortar as indicated by cohesion failure in the grout.

In a third test series mixtures of CaLoSiL® micro, CaLoSiL® E50, and HPC gel with different water content were applied to reconnect broken dolomitic lime mortar prisms (2 × 2 × 10 cm^3). Before treatment the mortar prisms had undergone a three-point bending strength measurement as described above. Subsequently, grouting mixtures were applied into the gap (six applications with syringes). The specimens cured 7 days after application at 20°C/ 65% RH. The three-point bending strength of the reconnected prisms was measured again The 1+1 mixture of CaLoSiL® E50 and CaLoSiL® micro with addition of HPC gel in ethanol and water developed a strength even higher as the previous strength of the prism (0.21 N/mm^2, see Fig. 5.10 no. 15). A higher content of CaLoSiL® micro or a lower content of HPC solution resulted in lower stability of the joint (see Fig. 5.10 no. 18, 19, 20). The mixtures of CaLoSiL® micro, CaLoSiL® E50, and HPC gel thus proved suitable for the consolidation of delaminations and cracks.

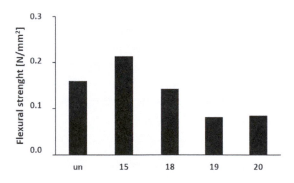

Figure 5.10 Three-point bending strength (N/mm^2) of reconnected broken mortar prisms. Mixtures tested: CaLoSiL® E50 + CaLoSiL® micro + thickener (mass/mass/volume): %15 = 1 + 1 + 0.4, 18 = 1 + 2 + 0.6, 19 = 1 + 3 + 0.8, 20 = 1 + 1 + 0.2 / un = untreated (single specimens).

5.3 Conclusion

Laboratory investigations focused on the improvement of distribution and retention of the calcium hydroxide nanosols. It was found that for structural consolidation of dolomitic lime mortar and gypsum mortar either dilution of the nanosols and stepwise application or after-treatment with cellulose ether gels resulted in strength increase of up to seven times of the untreated mortar. Even better results are possible by a combination of the single methods. The addition of a low amount of CaLoSiL® micro to CaLoSiL® nanosols improved the depth distribution of the consolidant and the mechanical strength of the treated mortar, too. Back-migration and formation of white haze could be avoided by these application procedures. In order to match specific conservation demands of deteriorated plaster, wall painting, and stucco work a 'module system' for modifying the nanosols was developed. The modified nanosols proved applicable for the consolidation of severely cracked limestone as well. For grouting and filling a mixture of CaLoSiL® and CaLoSiL® micronanosols can either be thickened with cellulose ether or can be filled. The test series demonstrate the feasibility of almost water-free lime-bound grouts with suitable properties. The achievable mechanical strength of the grouts is acceptable, although pure lime bonding reaches its natural limits. It cannot compete with hydraulically bound lime grouts for stronger consolidation treatments (e.g., large-scale reconnection of plaster). On the other hand the nanolime-based grouts open up new possibilities for the consolidation of fragile paint or plaster with water sensitivity. Still desirable are further studies in the ageing properties of conservation material based on calcium hydroxide nanosols and dispersions. Furthermore, efficiency in the presence of soluble salts has not yet been studied in detail.

Chapter 6

The Combination of Nanolime Dispersions with Silicic Acid Esters

Małgorzata Dobrzynska-Musiela,[a] Ewa Piaszczynski,[b] Elisabeth Mascha,[c] Jaroslav Valach,[d] Gerald Ziegenbalg,[e] and Claudia Dietze[e]

[a] Restauro Sp. z o.o., ul. Wola Zamkowa 6, 87100 Toruń, Poland
[b] Strotmann & Partner, Hauptstrasse 140, 53721 Siegburg, Germany
[c] University of applied Arts Vienna, Institute of Arts and Technology/Conservation Sciences, Salzgries 14, 1010 Vienna, Austria
[d] Institute of Theoretical and Applied Mechanics of the Czech Academy of Sciences, Prosecká 76, Prague 9, Czech Republic
[e] IBZ-Salzchemie GmbH & Co. KG, Schwarze Kiefern 4, 09633 Halsbrücke, Germany
malgorzata.musiela@restauro.pl; werkstatt@restaurierung-online.de;
liz.mascha1@gmail.com; valach@itam.cas.cz; gerald.ziegenbalg@ibz-freiberg.de;
claudia.dietze@ibz-freiberg.de

6.1 Introduction

Granular disintegration is a specific stone deterioration phenomenon that occurs mainly in sandstone, marble and granite. The composite grains become loose and detached. The following terms are often used to describe this particular form of stone

Nanomaterials in Architecture and Art Conservation
Gerald Ziegenbalg, Miloš Drdácký, Claudia Dietze, and Dirk Schuch
Copyright © 2018 Pan Stanford Publishing Pte. Ltd.
ISBN 978-981-4800-26-6 (Hardback), 978-0-429-42875-3 (eBook)
www.panstanford.com

(a)　　　　　　　　　　　　　　(b)

(c)

Figure 6.1　Typical examples of flaking and sanding stones.

decay [2, 135]:

- **Powdering, chalking:** terms sometimes used to describe the granular disintegration of finely grained stones
- **Sugaring:** employed mainly for white crystalline marble
- **Sanding:** used to describe the granular disintegration of sandstone and granite
- **Flaking:** scaling in thin, flat or curved scales of a submillimetre thickness

Typical examples are shown in Fig. 6.1.

The consolidation of disintegrated granular materials is one of the most important tasks in conservation. The main challenge is to find materials that are able to penetrate deteriorated structures and induce processes within them that lead to enhanced mechanical properties. Over the past few decades, silicic acid esters (SAEs), especially tetraethyl orthosilicate (TEOS), have found widespread use in this field. Due to their chemical similarity to sandstone and simple application, they are one of the most commonly and successfully used materials in stone conservation. SAE-based products are available in different forms and can be divided up into the following classes:

- Pure SAE with or without any catalysts to accelerate the gel formation
- SAE mixed with different organic solvents
- Pre-hydrolysed SAE alone or in combination with organic solvents
- SAE with additional functional components to enhance binding on non-siliceous surfaces
- SAE containing fillers and/or components to reduce or prevent shrinkage

Their use on different types of stone as well as their compatibility and retreatability have been the subject of controversial discussions. Fundamentals of the application of SAE will be summarised in the following together with possible combinations with nanolime.

6.2 Fundamentals of the Application of SAEs

Esters are compounds that are obtained by a reaction between alcohols and acids. Silicic acid esters are formally the products of reactions between silicic acid (H_4SiO_4) and different alcohols. In large scale, they are produced by the solvolysis of silicon tetrachloride ($SiCl_4$) with alcohols or alcohol/water mixtures. The characteristic feature of silicic acid esters is the presence of Si-OR bonds, where R stands for organic residues such as CH_3, C_2H_5, etc.

Esters can be hydrolysed by a reaction with water to produce the corresponding alcohols and acids. Basically, this hydrolysis can be induced by acidic or alkaline conditions (or metal organic catalysts).

The most important silicic acid ester used in conservation is TEOS ($Si(OC_2H_5)_4$). This is a clear, low-viscosity liquid with a typical smell. It is sold as 'pure' TEOS, as a partially prehydrolysed mixture, or diluted with different types of organic solvents (alcohols, hydrocarbons, ketones, etc.). Selected physico-chemical properties of TEOS are summarised in Table 6.1, the structure is shown in Fig. 6.2.

Table 6.1 Physico-chemical properties of TEOS

Molecular weight	208.32 g/mol
Density	0.94 g/cm^3
Boiling temperature	168°C
Melting point	−82.15°C
Vapour pressure at 20°C	9.21 hPa

It is important to note that TEOS is immiscible with water. However, ternary solutions can be obtained by adding ethanol or other alcohols (see Fig. 6.3).

Under the conditions of stone or mortar consolidation, the hydrolysis of TEOS is induced by a reaction with humidity. The most important property of TEOS is that hydrolysis results in the formation of polymeric networks consisting of Si-O-Si bonds. Complex reactions lead to the release of ethanol and the formation of gel structures. The main stages are:

(1) Hydrolysis of Si-OR bonds through an attack by water molecules: acid-catalysed, hydroxide-ion promoted or metal organic-catalysed
(2) Condensation/polymerisation resulting in the formation of silicic acid gels
(3) Gel ageing associated with shrinkage and the release of solvents (water, ethanol) incorporated into the gel structures (called 'syneresis')

Figure 6.4 and 6.5 illustrate the general reaction mechanism for the acidic and alkaline-based hydrolysis and condensation of silicic acid esters [136]. The hydrolysis of silicic acid esters is a slow process

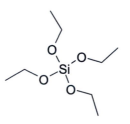

Figure 6.2 Chemical structure of TEOS.

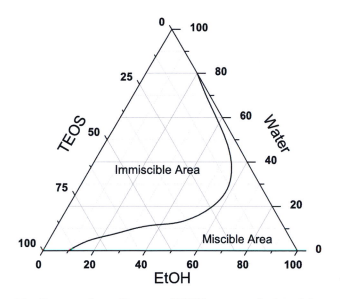

Figure 6.3 Ternary phase diagram of TEOS, water and ethanol (according to Ref. [135] and based on own measurements).

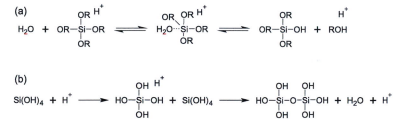

Figure 6.4 Reaction mechanism of (a) acid-catalysed hydrolysis and (b) condensation of silicic acid ester with a residue -R.

under normal conditions. Catalysts are added to speed up reaction and/or to influence and determine the properties of the resulting gels.

Gels formed under alkaline conditions have highly condensed structures compared to an acid-catalysed formation, which result in more linear and less branched structures.

The step in the acid- catalysed hydrolysis that determines the rate of reaction is the nucleophilic attack on the silicon atom by

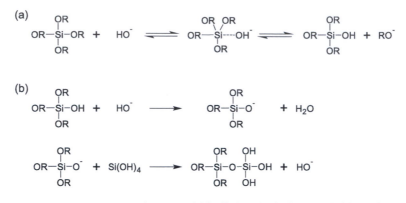

Figure 6.5 Reaction mechanism of (a) alkaline hydrolysis and (b) condensation of silicic acid ester with a residue -R.

the oxygen contained in a water molecule [137]. The alkoxide groups are initially protonated; hydrolysis then proceeds rapidly and is nearly complete before condensation starts [138]. For this to happen, an unprotonated $Si(OH)_4$ molecule reacts with a protonated one to create siloxane bonds. The catalyst is regenerated and condensation continues. However, it is important to know that monomers prefer to react with other monomers rather than with condensed silanols during acid-catalysed condensation. This is why entangled, linear or randomly branched chains are mostly formed under such conditions.

Hydrolysis under alkaline conditions involves a nucleophilic attack on the silicon atom by hydroxide anions (OH^-), followed by the displacement of an alkoxide anion (RO^-) [139, 140]. After rapid hydrolysis, condensation followed by deprotonation of a monomer and a subsequent reaction with $Si(OH)_4$ molecules creates a siloxane bond. The catalyst is regenerated and condensation continues. In this case, condensed silanols react with each other to produce highly branched structures.

Tin-organic compounds (tin dibutyl laureate compounds or related ones) are typical catalysts that were used in the past. Unfortunately, the composition of the catalysts used today are often not published by the producer. Tri-substituted organostannic compounds such as tributyltin and triphenyltin compounds in particular have been banned in the European Union since 2010 because of their high toxicity.

Stone consolidants on the basis of SAEs are characterised by their gel deposition potential. Pure TEOS has a gel deposition potential of around 30%, which theoretically results in the formation of about 300 g/L SiO_2. The resulting gel structures depend not only on the type of TEOS used, but also to a large extent on temperature, moisture and the catalyst that is used. It has to be remembered that the use of SAE leads to hydrophobic properties of the treated materials in the first step. Hydrophilic properties evolve in time and with an increasing degree of hydrolysis. A typical property of gels formed by TEOS hydrolysis is that water, ethanol and TEOS molecules are incorporated in the polymeric network. Water and ethanol evaporate over time and the gel 'dries'. This process is called 'syneresis' and results in shrinkage of the gel.

SAEs are particularly expedient for the consolidation of sandstone and sandstone-based materials. Due to the similar chemical character of the mineral surfaces, which exhibit Si-OH bonds, strong interactions can be achieved with the resulting gels. The application of SAE for clay-based materials as well as for calcareous minerals and materials has been the subject of controversial discussions.

The hydrolysis of TEOS within porous materials commonly results in brittle gel plates with sizes of approximately 10 μm. This means that conventional, TEOS-based stone consolidants are not suitable to fill larger pores or void structures. The resulting gels often have an undesirably high modulus of elasticity, which may result in an over-strengthening of treated zones.

Numerous publications have dealt with the problem of strengthening calcareous materials with silicic acid esters. The resulting silica gels often part from treated surfaces and no strengthening is achieved. There have been several attempts to overcome this problem. One possibility is seen in the formation of adhesive layers, for example based on calcium tartrate. To this end, the calcareous surfaces are treated with ammonium tartrate solutions. A hydroxylated surface forms onto which the SAE can anchor [141]. A different approach is the development of special admixtures of coupling agents to improve adhesion between the silica gel and calcareous surfaces. Aminosilanes are mainly used for this purpose. They can either be mixed directly with the SAE or applied separately as primer. The second method has often proved to be more promising.

The success of stone consolidation with SAE depends on both the characteristics of the material to be treated and the climatic conditions (temperature, humidity, wind, etc.) during the application. Mineralogical composition, salt content and the dimensions of the cracks, fissures or delaminations that have to be stabilised are very important as well.

The following aspects summarise the main drawbacks when using SAEs:

- Their use is irreversible, meaning that conservation treatments cannot be undone in the future.
- SAEs are unable to bridge large spaces, making them unsuitable as a consolidant for flaking and scaling stones.
- Problems occur when permanent hydrophobic properties remain after the application, especially where relative humidity conditions and the moisture content of the material due to sorption are either too low or too high.
- It is difficult to apply SAE-based consolidants on objects that contain a high concentration of salts and/or are affected by damp.

A comprehensive survey of properties, applications and the characteristics of materials treated by SAE can be found in Ref. [136].

6.3 Possibilities for the Combination of SAEs and Nanolime

To overcome some of drawbacks of the consolidation with SAEs, their combination with nanolime has been investigated. The combined application of nanolime and silicic acid esters is possible in two ways:

- As a homogeneous mixture
- By consecutive applications of nanolime dispersions and SAEs

The combination of both consolidants has the following advantages:

- The alkaline character of nanolime induces enhanced gel formation due to the presence of hydroxide ions. The higher the nanolime content, the faster gelation takes place and the higher the mechanical stability of the formed gels.
- Nanolime contains small amounts of water as a result of the way it is synthesised. Due to the presence of excess ethanol, *iso*-propanol or *n*-propanol, it can be mixed with an SAE in any ratio. Water also accelerates the hydrolysis of the SAE. If necessary, extra water can be added to the nanolime-SAE mixtures to support the mixing ratio and the reaction time.

6.3.1 Fundamental Laboratory Investigations to Characterise the Use of Homogeneous Mixtures of SAEs and Nanolime

Commercially available SAEs such as KSE 100 (Remmers, Germany) can be mixed with nanolime (calcium hydroxide nanosols) over a wide range. This produces homogeneous, opal-white solutions. Depending on the concentration of the nanosols, as well as the mixing ratio with SAEs, gel formation takes place between a few minutes and several days. Table 6.2 shows the time needed for gel formation depending on the mixing ratios between KSE 100 and

Table 6.2   Time needed for gel formation with different mixtures of CaLoSiL® E50 and E25 with KSE 100

Mixing ratio [vol%]	Substance	Gel time
1 : 1	E50 : KSE 100	2 h
1 : 2	E50 : KSE 100	4 h
1 : 5	E50 : KSE 100	24 h
1 : 10	E50 : KSE 100	36 h
1 : 1	E25 : KSE 100	15 h
1 : 2	E25 : KSE 100	15 h
1 : 5	E25 : KSE 100	24 h
1 : 10	E25 : KSE 100	36 h

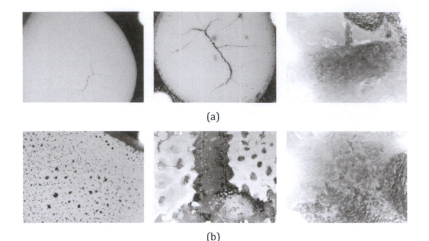

(a)

(b)

Figure 6.6 Microscopic pictures of films formed after the application of (a) CaLoSiL® E50 and KSE 100 mixtures (CaLoSiL® : KSE 1 : 1; 1 : 2; 1 : 5 vol%, from left to right); (b) CaLoSiL® E25 and KSE 100 mixtures (CaLoSiL® : KSE 1 : 1; 1 : 2; 1 : 5 vol%, from left to right) on glass surfaces. The films were allowed to dry for one week.

nanolime. It is evident that higher nanolime concentrations result in faster gel formation.

Applying nanolime-SAE mixtures to impermeable surfaces produces homogeneous films. The character of the films depends on the SAE and the nanolime dispersions used, the mixing ratio and the surface characteristics of the material to be treated. Films that formed after the application of CaLoSiL® (E50, E25) and KSE 100 mixtures are shown in Fig. 6.6. Ideally, homogeneous and dense films that adhere strongly to the substrate are obtained. Phase separation, crack formation or spalling do not take place in favourable mixtures.

The risk of cracks and spalling increases if high CaLoSiL® concentrations are used. This is due to fast gel formation. Nevertheless, the film formation also depends on the SAE used. Figure 6.7 shows different commercially available SAEs in combination with CaLoSiL® E25.

Calcium hydroxide initiates and accelerates the hydrolysis of the SAE, resulting in the formation of amorphous gels. Comprehensive

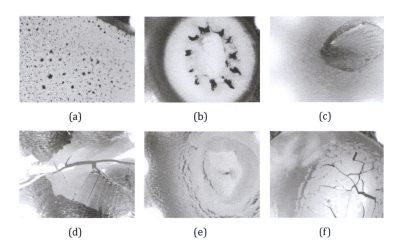

(a)   (b)   (c)

(d)   (e)   (f)

Figure 6.7 Microscopic pictures of films formed after the application of CaLoSiL® E25 and different SAE products (1:1 vol%) on a glass surface (magnification 20x). Drops were allowed to dry for one week. (a) CaLoSiL® E25:KSE 100, (b) CaLoSiL® E25:KSE 300 (Remmers), (c) CaLoSiL® E25:Protectosil® SH (Evonik, Germany), (d) CaLoSiL® E25:Dynasylan® A (Evonik), (e) CaLoSiL® E25:TEOS (Merck, Germany), (f) CaLoSiL® E25:Silres® BS OH 100 (Wacker, Germany).

XRD investigations did not produce any indications of the formation of crystalline calcium silicates. Figure 6.8 shows scanning electron microscopy (SEM) pictures of CaLoSiL® E25 mixed with KSE 100 on a glass surface.

The mechanical properties of mortar prisms after treatment with homogeneous CaLoSiL®-SAE mixtures were determined to assess the consolidation effect. Mortar prisms with dimensions of 4 cm × 4 cm × 4 cm were prepared by mixing lime hydrate and sand at a ratio of 1:6 by weight. All prisms were stored for one month until complete carbonation had been achieved. The results are summarised in Table 6.3. The tests were performed as follows:

(1) Saturation of the prisms with the CaLoSiL®-SAE mixture
(2) 48 h storage
(3) Second application of CaLoSiL®-SAE mixtures
(4) A final application of the mixture 48 h later
(5) Two month storage at 60% relative air humidity
(6) Determination of the compressive strength

Table 6.3 Compressive strength of mortar prisms (lime hydrate : sand = 1 : 6 by weight) after treatment with CaLoSiL®-SAE mixtures

Mixture	Number of applications	Compressive strength [N/mm^2]
CaLoSiL® IP25 :	1	0.4
KSE 300	2	0.7
(vol% 5 : 1)	3	1.2
CaLoSiL® E50 :	1	0.5
KSE 300	2	0.9
(vol% 5 : 1)	3	1.6
Reference	–	0.2

Summing up, the tests showed that a consolidation with nanolime-SAE mixtures is possible. According to Table 6.3, the compressive strength could be improved two- to eight-fold depending on the number of applications and mixture.

6.3.2 Fundamental Laboratory Investigations to Characterise the Consecutive Application of Nanolime and SAEs

Instead of using homogeneous mixtures of CaLoSiL® and SAEs, a successive application of both components is possible. However, it is important that nanolime dispersions be applied in the first step. The SAEs are applied after the alcohol has evaporated. Applying

(a)   (b)

Figure 6.8 SEM micrographs of CaLoSiL® E25 mixed with KSE 100 on a glass surface (1 : 1 vol%).

nanolime dispersions after the use of SAEs does not have any positive effects. Pre-treatment with SAE will saturate and seal the pore system of the stone and a subsequent CaLoSiL® application is ineffective.

The following tests were performed to characterise the consecutive application of different types of nanolime with different, commercially available silicic acid esters:

- Characterisation of films formed on glass surfaces after pre-treatment with nanolime dispersions followed by the SAE.
- Determination of the mechanical properties of mortar samples after several treatments.
- Consecutive saturation of cylinders filled with loose stone powders ('dummy' samples) with the consolidants and determination of the mechanical properties of the resulting bodies. This was realised either by dripping the consolidants onto the surface of the powders using a pipette or by capillary suction (see Fig. 6.9a).

(a)

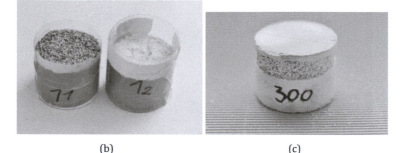

(b)   (c)

Figure 6.9 Samples used for characterisation. (a) Dummy sample of stone powder. (b) Cylindrical sample. (c) Sandwich sample.

- Simulation of sanding surfaces by covering a cylindrical stone piece with up to 2 mm of finely powdered stone dust and consolidation of the loose zone by repeated treatment with the consolidants until saturation. The samples were then stored in a humid environment (75% RH). Both the strengthening of the loose material and the connection to the intact stone were characterised (Fig. 6.9b).
- In order to characterise consolidation possibilities for loose zones between dense surface layers and intact stone, so-called 'sandwich' samples were developed. In these, stone powder was introduced between two pieces of intact stone. This 'disintegrated zone' was then treated by injections with syringes. Measurements of the drilling resistance of the treated zone, the adhesive tensile strength as well as microscopic methods were used to characterise the strengthening that was achieved (Fig. 6.9c).

Figure 6.10 illustrates the film formation after a consecutive application of CaLoSiL® and KSE 100 on glass slides. The ester was applied 1 h, 24 h and 72 h after the nanolime dispersion.

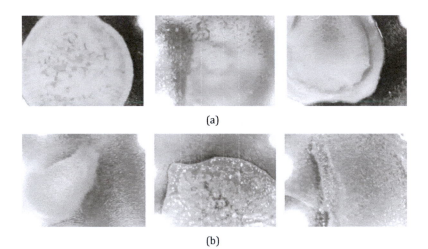

(a)

(b)

Figure 6.10 Consecutive application of (a) CaLoSiL® E50 and KSE 100 (applied 1 h, 24 h, 72 h after nanolime dispersion, from left to right); (b) CaLoSiL® E25 and KSE 100 (applied 1 h, 24 h, 72 h after nanolime dispersion, from left to right) on a glass slide (magnification 20x).

(a)   (b)   (c)

(d)   (e)   (f)

Figure 6.11 Penetration of KSE 100 and KSE 300. After consolidation with CaLoSiL® E25. (a–c) quartz sand, (d–f) red marble.

The pictures show that a consecutive application of CaLoSiL® and SAE is only useful if the respective SAE is applied between 1 h and 24 h after the nanolime. After this, no crack formation or other irregularities are evident. A rest time of more than 24 h leads to phase separation and the formation of inhomogeneous films (Fig. 6.10b, middle, right). As a follow up to these results, it was investigated how a pre-treatment with nanolime affects the migration of silicic acid esters into stone or mortar (Fig. 6.11). Samples consisting of quartz sand (grain size 0 to 1 mm) and red marble aggregate, 'Alpen Rosa' (grain size 0 to 8 mm) by Monte-Novo, were prepared to investigate this. They were filled into plastic rings with a diameter of 5 cm and a height of 5.5 cm. The samples were pre-consolidated with CaLoSiL® E25. After evaporation of the alcohol, the samples were placed in petri dishes. The bottom of each petri dish was filled with different SAEs (applied by syringes) and the capillary suction into the samples was observed. All of the tests showed that there is no risk of densification by pre-treatment with CaLoSiL®. The SAE penetrated the samples without any problems,

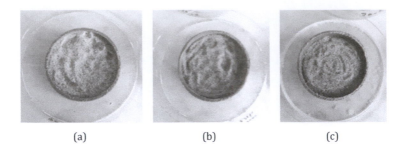

(a)   (b)   (c)

Figure 6.12   Sea sand treated with CaLoSiL® E25 and KSE 100 after (a) 1 h, (b) 24 h and (c) 72 h.

even when the samples had been pre-treated with CaLoSiL® up to eight times.

Following the positive capillary suction tests with both consolidants, the consolidation of loose sea sand was also tested. The sand was hereby placed into O-rings that were stuck to a petri dish. The sand was treated with CaLoSiL® E25 in the first cycle. KSE 100 was applied after 1 h, 24 h and 72 h (Fig. 6.12). The sample treated with KSE 100 after 1 h was hydrophilic after one week, unlike the other two samples, which remained hydrophobic for more than one month. The highest strength was achieved following the application of KSE 100 after 24 h. The entire sand sample was strengthened in this case. Treatment with KSE 100 after 1 h or 72 h resulted in surface stabilisation only, the sand core of the sample was still loose.

The tests were repeated with mortar prisms to determine the compressive strength that can be achieved. The results are summarised in Table 6.4. All prisms were prepared as follows:

- White lime hydrate : sand mixtures at a ratio of 1 : 6 by weight
- Storage for 28 days under room conditions at 80% RH

The mortar samples were saturated once with different CaLoSiL® products and 1 h, 24 h and 72 h later with KSE 100.

The data clearly indicates that the highest compressive strengths are achieved when the treatment with KSE 100 is carried out 24 h after the application of the CaLoSiL® products. Treatments 1 h or 72 h after the application of CaLoSiL® result in samples with a lower compressive strength. It is also clear that the maximum strength is already achieved after 14 days storage. This is confirmed by

Table 6.4 Consolidation of mortar prisms by consecutive treatment with CaLoSiL® E25 and KSE 100 measured after 7, 14 and 28 days ($n = 3$)

CaLoSiL®	KSE 100 applied after	Compressive strength [N/mm²] after 7 days	Compressive strength [N/mm²] after 14 days	Compressive strength [N/mm²] after 28 days
E50	1 h	0.5 ± 0.1	0.4 ± 0.1	0.5 ± 0.2
E50	24 h	1.3 ± 0.2	1.5 ± 0.2	1.3 ± 0.2
E50	72 h	0.9 ± 0.1	1.0 ± 0.3	1.2 ± 0.1
E25	1 h	0.7 ± 0.4	0.7 ± 0.2	1.0 ± 0.3
E25	24 h	1.3 ± 0.06	1.6 ± 0.2	1.7 ± 0.4
E25	72 h	0.5 ± 0.3	0.9 ± 0.1	0.7 ± 0.2
IP25	1 h	0.8 ± 0.04	0.6 ± 0.4	0.9 ± 0.5
IP25	24 h	1.3 ± 0.1	1.5 ± 0.06	1.4 ± 0.1
IP25	72 h	0.6 ± 0.2	0.9 ± 0.2	0.7 ± 0.3
Reference			0.1 ± 0.04	

the hydrophilic character of the surface of the mortars. It can be assumed that 1 h after the application of CaLoSiL®, small amounts of the alcohols are still present in the sample. They probably protect the $Ca(OH)_2$ particles against attack from the silicic acid ester. A significant share of the calcium hydroxide nanoparticles that form has already been converted into calcium carbonate after 72 h. Thus, the catalysing and adhesion-promoting properties of the nanolime particles are lost. This can partly be prevented by using higher concentrations of nanolime dispersions. Pre-treatment with CaLoSiL® E50 results in a slight increase in compressive strength, even when the SAE is applied 72 h after treatment with nanolime. This indicates that sufficient unreacted nanolime was still present. Compared to the application of homogeneous nanolime-SAE mixtures, the consecutive application of nanolime and SAEs achieves greater strengthening [142].

6.3.2.1 Investigations to consolidate disintegrated natural stone: Römer tuff, Nordhessischer tuff and Baumberger sandstone

The consolidation and strengthening of Römer tuff, Nordhessischer tuff and Baumberger calcareous sandstone always poses a challenge

Figure 6.13 Sandwich samples. From left to right: Marble, Drachenfels trachyte, Gotland sandstone, marl, Römer tuff, Nordhessischer tuff.

due to their specific properties. Consolidation with SAEs is difficult and generally results in insufficient strengthening. The successive application of nanolime and SAE, however, offers new possibilities. It is important that the original porosity is only partially reduced so that the stone remains capillary active.

Table 6.5 indicates that the combination of nanolime with KSE 300 results in a higher compressive strength than with KSE 100. This is due to the higher SAE content in KSE 300. The results of the tests on stone prisms (Table 6.5) were confirmed by investigations on sandwich samples, in which loose aggregates of tuff, marl, sandstone, limestone and marble (Fig. 6.13) were treated six times with CaLoSiL® E25 followed by the application of KSE 100 and KSE 300. A significant increase in the adhesive strength was achieved in every case (Fig. 6.14).

Drilling resistance measurements in the consolidated zones (between the two compact stone pieces) also indicated a significant improvement in strength compared to the single application of SAE (Fig. 6.15).

SEM micrographs were taken to study the consolidated structure in sandwich samples. Typical pictures of marl and Drachenfels trachyte samples are shown in Fig. 6.16.

The structure and morphology of the resulting silica gels depend on the substrate as well as the SAE used and the nanolime content. Amorphous silica gels are normally formed in which calcium is homogeneously distributed. If nanolime has been used in high concentrations or in several cycles before the SAEs are applied, the $CaCO_3$ modifications calcite and vaterite have also been known to

Table 6.5 Characteristics of Römer tuff, Nordhessischer tuff and Baumberger sandstone

	Before consolidation	After consolidation (6x CaLoSiL® E25)	After consolidation (1x CaLoSiL® E25 + KSE 100)	After consolidation (1x CaLoSiL® E25 + KSE 300)
Römer tuff (weathered)				
Water absorption [wt%]	23.24	25.89	19.5	17.9
Porosity [vol%]	39.0	31.05	30.7	29.2
Compressive strength [N/mm²]	6.3	7.0	7.8	8.5
Modulus of elasticity [N/mm²]	4000	4000	4230	4500
Hygric elongation [mm]	1.4	1.41	1.45	1.49
Nordhessischer tuff (weathered)				
Water absorption [wt%]	16.5	18.3	15.4	13.89
Porosity [vol%]	43.61	45.9	39.66	37.99
Compressive strength [N/mm²]	17	17.9	18.08	19.6
Dyne Young's modulus [N/mm²]	9000	9000	n.d.	n.d.
Hygric elongation [mm]	0.4	0.42	0.42	0.421
Baumberger sandstone (weathered)				
Porosity [vol%]	19.08	20.4	18.45	18.2
Compressive strength [N/mm²]	41	42.1	43.1	45.1
Dyne Young's modulus [N/mm²]	6000	n.d.	n.d.	6444
Hygric elongation [mm/m]	0.2	0.21	0.211	0.225

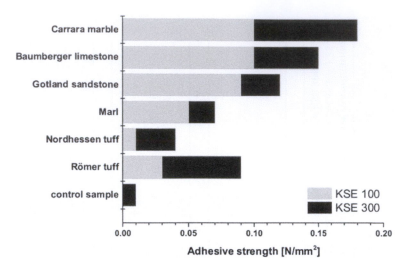

Figure 6.14 Adhesive strength of different stone samples consolidated with CaLoSiL® E25 in combination with KSE 100 and KSE 300.

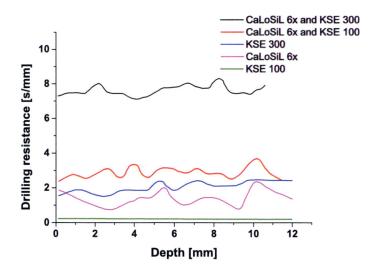

Figure 6.15 Drilling resistance measurement of CaLoSiL® E25, KSE 100 and KSE 300 alone and in combination treated marble 0–2 mm samples.

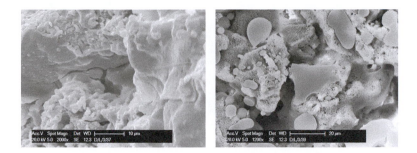

(a)   (b)

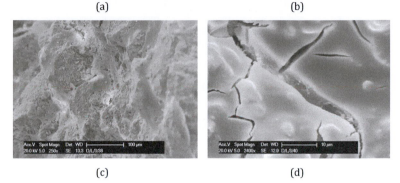

(c)   (d)

Figure 6.16   SEM micrographs of CaLoSiL® E25 with (a) KSE 100 and (c) KSE 300 in Marl as well as CaLoSiL® E25 with (b) KSE 100 and (d) KSE 300 in Drachenfels trachyte. Photo: E. Mascha.

form. They probably formed before the SAEs were applied. Typical structures obtained by various treatments are shown in Fig. 6.17.

The application of SAEs under high humidity is always difficult, and often impossible. Figure 6.18 shows samples that were only treated with KSE 100 and stored at 90% RH. No consolidation took place under such conditions. Samples that were prepared by consecutive saturation with CaLoSiL® E25 and KSE 100 and then stored for three weeks at 90% RH, however, are intact and show no signs of deterioration (see Fig. 6.19). The number of pre-treatments with CaLoSiL® depends on the characteristics of the aggregates and their particle sizes. Up to six treatments with CaLoSiL® did not show any negative effects on the migration of SAEs into the samples.

The successive application of nanolime dispersions and SAEs on marble aggregates results in samples characterised by a high

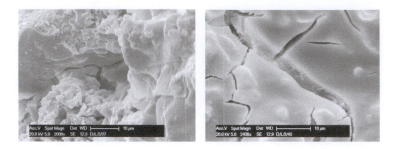

(a) (b)

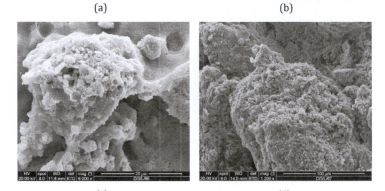

(c) (d)

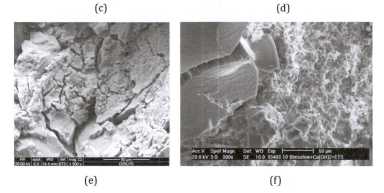

(e) (f)

Figure 6.17 SEM micrographs of (a) Marl stone powder treated with CaLoSiL® E25 followed by KSE 100. (b) Stone powder from Drachenfels trachyte treated with CaLoSiL® E25 and KSE 300. (c) Marl stone powder after two treatments with CaLoSiL® E25 followed by KSE 300. (d) Baumberger sandstone powder after two treatments with CaLoSiL® E25 followed by KSE 100. (e) Carrara marble powder after two treatments with CaLoSiL® E25 followed by KSE 300. (f) Pumice after the treatment with CaLoSiL® E25 and the Wacker BS-OH 100. Photos: E. Mascha.

Figure 6.18 Consolidation with KSE 100, one cycle. Top: Marble, Baumberger sandstone, Römer tuff, Drachenfels trachyte. Bottom: Marl, Kaspar yellow marble, quartz sand.

freeze-thaw resistance as well as a good resistance to salt crystallisation. The tests are based on freezing at $-17.5°C$ and subsequent thawing up to $20°C$. In the salt crystallisation tests, the samples were stored for 2 h in a saturated Na_2SO_4 solution and then dried at $105 \pm 5°C$. As the data in Table 6.6 indicates, the best results are obtained if the marble aggregates are treated with the silicic acid ester after one day. Using this procedure leads to a high compressive strength and minimum weight loss during freeze-thaw and salt crystallisation tests. Both an immediate (after 2 h) and later (after 6 days) second treatment produces samples with a lower compressive strength as well as a lower resistance to freeze-thaw and crystallisation stresses.

6.3.2.2 Investigations to consolidate disintegrated natural stone: Żerkowice sandstone, Gotland sandstone and Pińczów limestone

Cubes with edges of 50 mm were prepared to characterise the combined application of nanolime and SAE for the consolidation

Table 6.6 Results of frost-thaw cycles (DIN 52104) and salt crystallisation tests (DIN 52111) on marble aggregates 0–2 mm

Consolidant	Compressive strength [N/mm^2]	Freeze-thaw-cycle, weight loss after 8 cycles [%]	Salt crystallisation test, weight loss after 4 cycles [%]
Control sample: KSE 100	0.01	1 Cycle 100% 3 Cycles 100%	1 Cycle 100% 3 Cycles 100%
Control sample: KSE 300	0.15		
1 x CaLoSiL® E25, after 2 h KSE 100	0.02	36	43
1 x CaLoSiL® E25, after 2 h KSE 300	0.02	23	35
6 x CaLoSiL® E25, after 1 day KSE 100	0.08–0.11	10.3	15
6 x CaLoSiL® E25, after 1 day KSE 300	0.24	7.6	13
8 x CaLoSiL® E25, after 1 day KSE 100	0.2	10.9	12
8 x CaLoSiL® E25, after 1 day KSE 300	0.26	5.6	10
6 x CaLoSiL® E25, after 6 days KSE 100	0.05	28	34
6 x CaLoSiL® E25, after 6 days KSE 300	0.1	12	25
Addition of water, after 1 day, 6 x CaLoSiL® E25, after 1 day KSE 100	0.17	9	14
Addition of water, after 1 day 6 x CaLoSiL® E25, after 1 Day KSE 300	0.28	11	12

Figure 6.19 Consolidation with CaLoSiL® E25 and KSE 100 once. Top: Marble, Baumberger sandstone, Römer tuff, Drachenfels trachyte. Bottom: Marl, Kaspar yellow marble, quartz sand.

of Żerkowice sandstone, Gotland sandstone and Pińczów limestone. Holes with a diameter of 20 mm and depths between 10 and 20 mm were drilled in the middle of the cubes (Fig. 6.20). These were filled with sieved stone powder with particle sizes less than 0.12 mm. The powder was slightly compacted at the level of 2 mm beneath the surface of each shape.

The following treatments were realised:

- Saturation with CaLoSiL® E25 or CaLoSiL® E45, a second treatment was realised after 24 h
- Saturation with KSE 300
- Second saturation with CaLoSiL® E25 or CaLoSiL® E45 followed by the treatment with KSE 300 after 5 days

A volume of around $10 \, \text{cm}^3$ of CaLoSiL® or KSE was necessary to saturate the stone powder. In a second set of experiments, the powdered stone was moistened with water before packing. After a proper period of carbonation and conditioning (a total of 6 weeks), the samples were dry-cut parallel to the drilled holes using a

Figure 6.20 Model samples. Row on the left: hollowed Pińczów limestone cubes; middle row: Żerkowice sandstone; row on the right: Gotland sandstone.

diamond saw. They were then examined for the first time and tested for hardness with a needle to assess the level of the aggregate 'cementation'. The condition of the samples was documented in photographs (Fig. 6.21).

Comprehensive characterisation of the consolidated zones was realised by SEM. To this end, the entire sample, including the intact stone, were vacuum-embedded in epoxy resin (Araldite® 2020). Polished cross sections were then produced perpendicular to the surface of treatment. These were coated with carbon and examined under an SEM (Philips XL 30 ESEM, 20 KV, high vacuum, back-scattered electron [BSE] detector) as well as in an energy-dispersive X-ray analyser (Link-ISIS). The SEM micrographs taken at a low magnification were joined using Photoshop® so as to cover the entire sample diameter. Pores were edited in pseudo colour (blue) to make them more conspicuous and to allow a comparison of the different treatment methods (Fig. 6.22).

The qualitative evaluation of the consolidation effect that was achieved is summarised in Table 6.7.

From the point of view of a microstructure examination, the best consolidant distribution in the reinforced material was achieved

Table 6.7 Summary of microscopic study results

	Consolidant in stone	Layer on surface	Accumulation of fine aggregates on border	Distribution of consolidant	General assessment
Żerkowice sandstone + SAE	Yes	No	Yes	Very bad	−
Żerkowice sandstone + CaLoSiL®	None	Yes	Yes	Good	++
Żerkowice sandstone + CaLoSiL® + SAE	None	Yes	Yes	Very good	+++
Gotland sandstone + SAE	Yes	No	Yes	Very bad	−
Gotland sandstone + CaLoSiL®	None	Yes	Middle	Good	++
Gotland sandstone + CaLoSiL® + SAE	None	Yes	Middle	Good	++
Pińczów limestone + SAE	None	No	Middle	Good	−/+
Pińczów limestone + CaLoSiL®	None	Yes	Middle	Very good	+++
Pińczów limestone + CaLoSiL® + SAE	None	Yes	Middle	Very good	+++

Figure 6.21 The samples were cut in half after 6 weeks of seasoning. The figure shows the various effects of consolidation treatment.

with Pińczów limestone and Żerkowice sandstone treated with a SAE and nanolime in the sequence CaLoSiL® E45 first and SAE second. It is worth noting that SAE, commonly used in conservation treatment, did not prove to be at all effective as a single agent for aggregate consolidation. Little penetration was found in Żerkowice and Gotland sandstone samples imitating deteriorated stone, whereas it was clearly present in the 'healthy' stone fragments (see Fig. 6.22). The most noticeable influence of nanolime could be observed during the consolidation of weak and deteriorated materials, both natural stones and mortars. The SEM micrograph in Fig. 6.23 supports this finding and demonstrates a strengthening

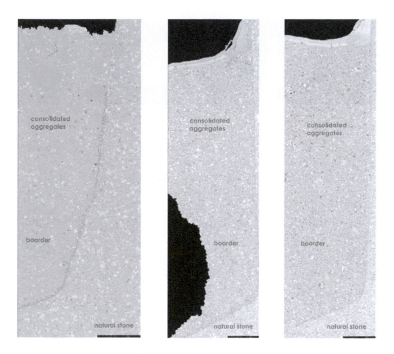

Figure 6.22 Gotland sandstone samples consolidated with different agents and their combinations: SAE on the left, nanolime in the middle, nanolime and SAE on the right. A homogeneous consolidation effect and a good merging of the aggregate filling with the healthy stone were only achieved when both agents were applied. Photos: E. Mascha.

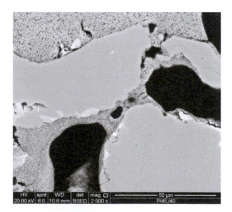

Figure 6.23 Electron microscopic image of a sample reinforced with nanolime. The lime formed consolidating bridges between loose grains. Photo: E. Mascha.

(a)   (b)

Figure 6.24 The figure shows samples, that were (a) first saturated with CaLoSiL® E45 with the aggregate moist, and then with SAE. (b) The consolidation effect remained unchanged after 20 freeze-thaw cycles.

of the loose limestone aggregate by bridges formed by nanolime particles.

The durability of the resulting aggregate consolidation was tested by freeze-thaw cycles. The tests were conducted under the following conditions:

- 6 h saturation with water through capillary action
- 18 h freezing $-20°C$

The samples were photographed before the tests and after 5, 10, 15 and 20 cycles. Tables 6.8 and 6.9 summarise the condition of the samples after 10 and 20 freeze-thaw cycles (Fig. 6.24).

The tests demonstrated that the most durable binding effect is achieved with CaLoSiL® E45 applied to a wet aggregate with the secondary introduction of SAE (KSE 300). After 20 freeze-thaw

Table 6.8 The condition of the samples after 10 freeze-thaw cycles

Impregnant	Aggregate	Żerkowice sandstone	Gotland sandstone	Pińczów limestone
CaLoSiL® E45 + SAE	Dry	Minor depletion	Partially poured out	Partially poured out
CaLoSiL® E45	Dry	Partially poured out	Poured out	Poured out
CaLoSiL® E45 + SAE	Wet	No change	No change	No change
CaLoSiL® E45	Wet	Minor depletion	Partially poured out	Partially poured out
SAE	Dry	Partially poured out	Porous	No change

Table 6.9 The condition of the samples after 20 freeze-thaw cycles

Impregnant	Aggregate	Żerkowice sandstone	Gotland sandstone	Pińczów limestone
CaLoSiL® E45 + SAE	Dry	Minor depletion	Minor depletion	Minor depletion
CaLoSiL® E45	Dry	Partially poured out	Completely poured out	Poured out
CaLoSiL® E45 + SAE	Wet	No change	No change	No change
CaLoSiL® E45	Wet	Minor depletion	Poured out	Partially poured out
SAE	Dry	Major depletion	Porous substance	No change

cycles, the bound aggregates did not pour or fall out of the hole. Aggregates of all three stones could be successfully consolidated. The consolidation effect was slightly weaker when the same composition of impregnates was applied to a dry aggregate. Minor depletions in the bound aggregate were observed after 20 cycles.

If an aggregate, dry or wet, was treated only with nanolime, the samples began pouring off after 5 cycles (Pińczów limestone) and after 10 cycles (Gotland, Żerkowice sandstones). If bound solely with silicic acid ester, only samples of Pińczów limestone demonstrated a lasting consolidation effect. The two sandstones began to deteriorate after 5 cycles.

The drilling resistance was investigated in two different ways. In the first set of tests, holes were drilled to various depths in the consolidated aggregates and compared to holes drilled into intact stone. In a second series of tests, the drilling resistance was evaluated in a profile along the axis of the hole filled with aggregate for the same set of stones and treatments. The results for the first batch are summarised in Fig. 6.25.

It can be deduced that consolidants are more effective in the Pińczów limestone. If the three treatments are compared, the greatest increase in drilling resistance can be found in for the combined CaLoSiL® E45 and SAE treatment.

In the case of Żerkowice and Gotland sandstone, the intact material is so strong that it cannot be penetrated with the given drill force. The effect of consolidants on the drilling resistance of

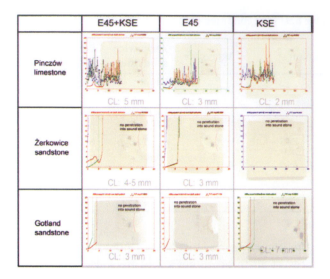

Figure 6.25 Drilling resistance plots for the studied aggregates in a direction perpendicular to the axis of the hole. Rows refer to stone type, columns to treatment procedure. The locations of drilled holes in the specimens are shown behind the drilling resistance plots; blue, green, red and black colours correspond to drilling resistance plot for hole 1, 2, 3 and 4, respectively.

Żerkowice and Gotland sandstones is very slight and the aggregates remain weak after treatment. There are two reasons for this finding: first, the particles in the aggregate are larger, so there are not many contacts where bridges can develop; second, the binder is based on a different chemistry. Some of the hole drilling records have also been lost because of the material was too loose or the aggregate became detached from its surroundings (Żerkowice sandstone in the case of SAE treatment).

As for the second batch, the analysis (Fig. 6.26) of drilling records suggests several important facts about the material under examination:

- There is noticeable increase in the drilling resistance as drilling progresses, even through homogeneous material. This is caused by the accumulation of removed material along the drill so that part of the drill's power is dissipated by friction.
- Material inhomogeneities of up to 0.1 mm can be seen in the plot.

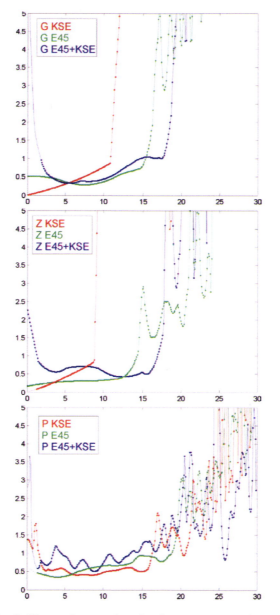

Figure 6.26 Drilling resistance plots for the aggregates under examination (top) Gotland sandstone, (middle) Żerkowice sandstone and (bottom) Pińczów limestone after various treatments ((red) SAE, (green) E45 and (blue) E45 + SAE). The horizontal axis depicts depth [mm] and the vertical axis drilling resistance [s/mm].

The development of a nanolime crust can also be seen in the case of a combined E45 and SAE treatment for both sandstones. As for the Pińczów limestone, the strengthening effect is noticeable over the entire drilled depth. In this case, there is no crust on the surface because the consolidant has mixed with the aggregate.

In short, the trials using CaLoSiL® E25 and E45 combined with SAE as a secondary binder for disintegrated stone materials proved effective, especially for Żerkowice sandstone and Pińczów limestone. The weakest consolidation effect was achieved with Gotland sandstone for all of the compositions used. It should be emphasised that the efficacy of the procedure was greatest if the aggregate had been slightly moistened with water beforehand. A combination of two impregnates worked much better than CaLoSiL® E25, E45 or KSE 300 on their own.

Chapter 7

Practical Applications

Karol Bayer,[a] Dana Macounová,[a] Małgorzata Dobrzynska-Musiela,[b] Luboš Machačko,[a] Ewa Piaszczynski,[c] Nadine Wilhelm,[c] Verena Wolf,[c] Arnulf Dähne,[d] Christoph Herm,[e] Thomas Köberle,[e] Zuzana Slížková,[f] Jan Vojtechovsky,[a] and Radek Ševčík[b]

[a] University of Pardubice, Faculty of Restoration, Jiráskova 3, 57001 Litomyšl, Czech Republic
[b] Restauro Sp. z o o., ul. Wola Zamkowa 6, 87100 Toruń, Poland
[c] Strotmann und Partner, Hauptstrasse 140, 53721 Siegburg, Germany
[d] pons asini, Schloss 17, 04600 Altenburg, Germany
[e] Hochschule für Bildende Künste Dresden, Güntzstrasse 34, 01307 Dresden, Germany
[f] Institute of Theoretical and Applied Mechanics of the Czech Academy of Sciences, Prosecká 76, Prague 9, Czech Republic
karol.bayer@upce.cz; macounova@itam.cas.cz; malgorzata.musiela@restauro.pl; lubos.machacko@upce.cz; karol.bayer@upce.cz; werkstatt@restaurierung-online.de; arnulf.daehne@pons-asini.de; herm@serv1.hfbk-dresden.de; slizkova@itam.cas.cz; jan.vojtechovsky@upce.cz

7.1 Introduction

In the STONECORE project, scientists and restorers from six European countries cooperated. The basic idea was to combine fundamental and applied research. During this project it was possible to treat extremely versatile objects ranging from ancient wall paintings in the Herculaneum (Italy) or the treatment of fissures and cracks in the ancient theatre of Megalopolis (Greece) to concrete

Nanomaterials in Architecture and Art Conservation
Gerald Ziegenbalg, Miloš Drdácký, Claudia Dietze, and Dirk Schuch
Copyright © 2018 Pan Stanford Publishing Pte. Ltd.
ISBN 978-981-4800-26-6 (Hardback), 978-0-429-42875-3 (eBook)
www.panstanford.com

conservation on the St. Martin church in Dahlem-Schmidtheim (Germany) with different nanolime dispersion, application methods and conservation strategies. Preliminary studies were performed on representative samples exhibiting all characteristic damages of the original object. Furthermore comprehensive damage analysis has been carried out as well as the development and realisation of specific conservation concepts for each object. The following sections summarise these investigations and are aimed to give an overview of the complexity of such treatments.

The special feature of this part of the book is the detailed description of conservation processes of 13 different historical objects by different authors/restorers. Each chapter represents the individual experiences, assessments as well as recommendations for the use of nanolime dispersions. All authors are available for further discussions. In direct contact, they will be able to provide much more detailed information than that can be given within the scope of this book. The corresponding addresses are given in each chapter. The authors hope that the following chapters support the information given in the previous chapters and provide an inside view in the various possibilities given for the application of nanolime dispersions.

7.2 Kutná Hora Angel*

7.2.1 Object: Basic Description

The Baroque statue of an angel with child, probably depicting Archangel Raphael with Tobias (from 1764), is a part of the Marian decorations of the outer wall of a town house in the centre of Kutná Hora – a set of three statues (Virgin Mary with two angels at the sides) (see Fig. 7.1). The statue was sculpted from a local porous limestone called Kutná Hora limestone, which has historically been used in this region for sculptural works. Originally the statue was decorated by a polychromy. The size of the statue is as follows: height 147 cm, the statue base 83 cm × 48 cm [143].

7.2.2 Stone Characterisation

As mentioned above, the sculpture is made of local limestone (calcarenite) so-called Kutná Hora limestone. This can be characterised

---

*This section is written by Karol Bayer and Dana Macounov.

(a)   (b)

Figure 7.1 (a) Outer wall of a town house in Kutná Hora with Marian sculptural decorations. (b) Statue of the angel on the wall before restoration.

as coarse to very coarse sandy biodetritic limestone. It belongs to the Mesozoic, Upper Cretaceous sediments of the Czech Cretaceous Basin Kutná Hora. Unusual round shape of individual grains of limestone (bioclasts) testifies to its origin from the upper sedimentary layers, which occurred in the waters of the Upper Cretaceous sea breakers for their processing, so it is also called the intertidal Kutná Hora shell limestone. The material is mineralogically inhomogeneous, composed of quartz grains, fragments of rocks, mica, and especially of the so-called biodetrit, which in this case consists mainly of residues of mollusc shells, especially oysters [143, 144]. The microscopic study (Fig. 7.2) of the non-weathered limestone, shows that the rock has a high porosity and the individual particles are contact-bonded by calcite. Pores constitute approx. 20 to 30% of the limestone volume and the pore size distribution is rather broad with pore size of the majority ranging from 1 to 2 μm up to 200 to 300 μm. High porosity and weak bonding of individual grains, explains the good workability of the material by carving, but on the other hand also it's relatively low corrosion resistance, particularly in environments laden by acidic atmospheric pollutants [145].

7.2.3 Condition Assessment: Damage Phenomena

Comprehensive investigation of the statue was performed prior to conservation treatment, including a detailed condition assessment.

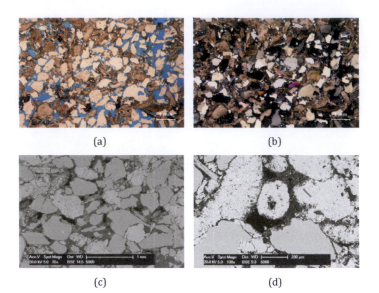

(a) (b)

(c) (d)

Figure 7.2 (a) Thin section of the non-weathered limestone, optical microscopy image, transmitted light, parallel nicols, the blue-coloured area of the pores are filled with blue-dyed epoxy casting resin. (b) Thin section of the non-weathered limestone, optical microscopy image, transmitted light, crossed nicols. (c) Scanning electron microscopy, back-scattered electron image (SEM-BEI); lightest particles are particles (fragments) of calcium carbonate (calcite), the grey particles are mostly quartz, the pores are dark grey to black. (d) SEM-BEI.

Generally can be stated the statue was in a very bad, nearly critical condition. The Kutná Hora limestone was in an advanced stage of degradation caused by several corrosion factors. Due to weathering there was a significant decrease in stone mass on the whole statue, locally up to tens of millimetres. The original surface including the polychromy of the statue was not protected against weathering during the previous storage preserved and a large area exposed to rainwater was covered with algae and locally with lichens and mosses. The statue was largely overgrown with ivy tendrils which were rooted into the limestone mass. Much of plastically distinctive shapes were repaired using a hard, dense repair cement mortar. The 'smooth' surfaces at the bottom of the statue were treated in the same way. The repair mortar overlapped at several places, altering the shape of the statue. Dark coatings and crusts covered the surface mainly on rain shadow areas. Underneath the crusts and repair mortar the stone was often heavily degraded and tended to sanding. The lower part of the statue was heavily stressed by rainwater as well as rising damp leading to a significant reduction in stone strength. In summary, the main areas of damage were as follows: rock surface corrosion, which caused the massive loss of stone mass; a surface covered by algae, mosses and lichens; ingrown ivy tendrils; extensive coverage of the surface and complementation of the missing parts by hard repair mortars based on Portland cement from the previous restoration and massive gypsum crusts, which were delaminating from the surface stone layers and deforming the statue surface. The following detailed photos and brief descriptions document the types of lesions located on the statue of the angel (Table 7.1) [143, 145].

7.2.4 Condition Assessment: Scientific Investigations

In order to investigate the basic properties of the stone and to clarify more precisely the condition of the statue, several non-invasive measurements and also different investigations were performed on samples taken at selected locations.

- Non-invasive (directly on the statue): determination of water absorption by Karsten tube, ultrasonic transmission, drilling resistance measurement (semi-invasive).
- Invasive (on samples): determination of water absorption porosity, bulk density, pore-size distribution, determination of the content of water soluble salts, microscopic analysis (optical microscopy and scanning electron Karsten) of the weathered stone, crusts, repair mortar and remnants of polychromy.

7.2.4.1 Determination of water and ethanol absorption by Karsten tube

The determination of water and ethanol absorption by Karsten tube was performed on three different areas of the stone surface with different damage phenomena (Figs. 7.3 and 7.4). According to the results in Table 7.2, the limestone was rated as highly absorbent.

Table 7.1 Photographs and brief descriptions of the types of lesions located on the statue of the angel

Damage	Position	Description
Surface erosion	Right side of the angels chest, detail	Gradual surface erosion caused mainly by 'wash out' of less resistant parts by acid rain leading to the loss of original material and roughening.
Biological colonisation	Hair on angels head, detail	The creation of a layer caused by biological colonisation (mainly algae, partly lichens and mosses).
Surface erosion and disintegration	Angels right hand, detail	Progressive stage of erosion of the limestone surface, causing rounding and significant loss of statue shape and its readability. Local deeper disintegration – sanding.
Formation of black crusts well connected with the stone surface	Left leg of the angel, detail	Formation of hard, compact gypsum crusts firmly connected to the stone underneath.
Formation of black crusts detaching from the stone surface; partly blistering	Right hand of the child, detail	Formation of hard, compact, black gypsum crusts detaching from the stone underneath; below the crust cracks and heavily disintegrated stone (sanding); local blister formation.

Table 7.1 (*Continued*)

Damage	Position	Description
Repair mortar (plastic retouching) detaching from the stone	Left knee of the angel, detail	Visually and physic-mechanically inappropriate repair mortar detaching with a thin layer of original limestone from the stone underneath. Below this repair mortar formation of cracks and heavily disintegrated stone (sanding).
Repair mortar (plastic retouching) well connected with the stone surface	Angels belt on the right side, detail	Visually and physic-mechanically inappropriate compact repair mortar well connected with the stone surface (without detachment).
Cracks	Base of the statue, detail	Formation of cracks caused by weathering or mechanical stress.
Missing parts	Base of the statue, detail	Larger parts of the statue broken off due to weathering or mechanical stress.

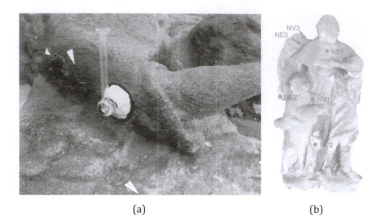

(a)   (b)

Figure 7.3 (a) Measurement by Karsten tube. (b) Plot of measurement points.

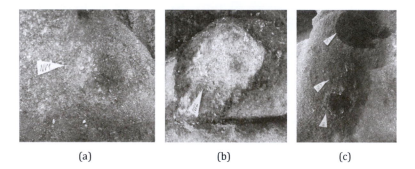

(a)   (b)   (c)

Figure 7.4 (a) Eroded surface with algae colonisation. (b) Eroded surface with massive loss of limestone. (c) Eroded surface covered by black crust.

High values demonstrate the ability of the stone to absorb fluids. This observation, along with the high porosity recorded, are important factors for the successful penetration of a consolidant. This also correlates well with the results of microscopic study.

A significant increase in absorption of fluids at the corroded areas also confirms the general assumption that the destruction of rocks associated with the deterioration of mechanical properties had been increasing their porosity. Slightly higher absorption measurements using ethanol can be explained by its lower surface tension and improved wetting of the surface stone. Biogenic coatings

Table 7.2 Measurement results – determination of water and ethanol absorption

Marking	Surface type	Medium	Water absorption coefficient W [$kg \cdot m^{-2} \cdot h^{-0.5}$]
NV1	Eroded surface	Water	22.8
NV2	Eroded surface	Water	106.3
NV3	Surface colonised with algae, fungi, etc.	Water	41.3
NE1	Eroded surface	Ethanol	31.6
NE2	Eroded surface	Ethanol	110.5
NE3	Surface colonised with algae, fungi, etc.	Ethanol	68.6

do not reduce water absorption surface of the stone, the higher absorbability probably explains the degradation of rocks beneath them.

7.2.4.2 Drilling resistance measurement

A drilling resistance investigation was carried out in order to find the conditions of the stone beneath the surface. The drilling resistance profile corresponds with the strength or hardness profile of the surveyed material. As this investigation is a 'semi-invasive' method it was only possible to perform measurements on a limited number of areas and on 'hidden' parts of the statue (Fig. 7.5).

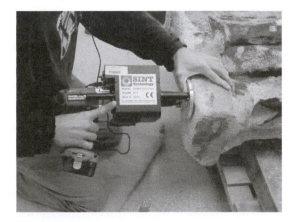

Figure 7.5 Drilling resistance investigation on the base of the statue.

Table 7.3 Characteristic results of DRMS measurement

Measurement spot description	DRMS-profile
R2 Bottom part of the base, 5 cm from the edge, fracture unexposed to weathering	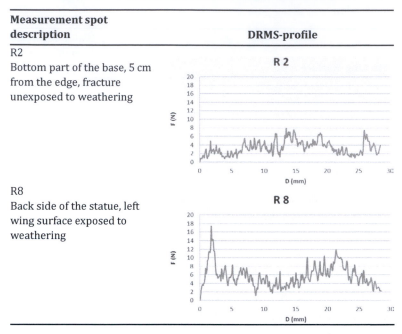
R8 Back side of the statue, left wing surface exposed to weathering	

The results of the non-weathered limestone (R2, Table 7.3) on the bottom part of the base shows a drilling resistance profile without any significant gradient and confirms the expected heterogeneity of limestone (the device used is quite sensitive and responsive to the presence of harder grains or mollusc-shells).

The profile for the exposed surface on the back of the angel (R8, Table 7.3) shows significantly higher surface strength (to a depth of 2–3 mm). This can be explained as a consequence of limestone surface sulphation, which frequently leads to the formation of more compact, less porous zones.

7.2.4.3 Ultrasound transmission measurement

The method is based on measurement of the ultrasonic waves passing through the material. The main evaluation criterion is the velocity of the ultrasound waves (p-waves). In addition, amplitude or signal shape can be used to assess condition of the statue.

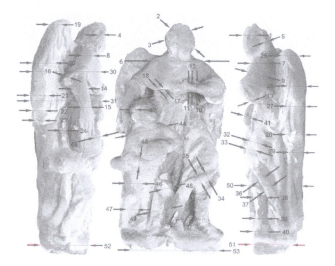

Figure 7.6 Plot of measurement points.

The aim of the measurements was to evaluate the extent of damage to the limestone, as well as to reveal potential hidden damages (e.g., voids and cracks) and to prepare data for objective comparison with the material condition after consolidation/conservation. A total of 53 measurements were carried out at various positions of the statues in order to represent different types of damage (Fig. 7.6 and Table 7.4).

The results (Fig. 7.7) can be summarised as follows:

- The average ultrasound velocity is relatively low at $v = 1.73$ km/s. The average velocity is caused on one hand by the properties of used limestone itself (e.g., high porosity, week contact-bonding of clastic particles) and on the other hand also significantly influenced by the extensive stone weathering. For this reason, a relatively large difference was found between

Table 7.4 Average, maximal and minimal ultra-sound-wave velocities

Average US-velocity	1.73 km/s
Minimal US-velocity	0.77 km/s
Maximal US-velocity	2.75 km/s

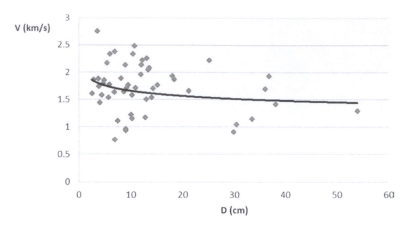

Figure 7.7 Profile of ultrasound-wave velocities (relationship between the US-velocity and the measurement distance).

the highest and lowest measured values (maximum US-velocity 2.75 km/s, the minimum US-velocity 0.77 km/s). The lowest values have been measured at places with obviously highest degree of limestone degradation.

- Under the crust and the repair mortars, there are often cracks or degraded zones. The repair mortar layers tend to detach from the underlying limestone.
- The base of the statue is also heavily damaged by limestone disintegration.
- There is an evident slightly increasing trend of US-velocity towards the surface in general. Similarly as in the evaluation of similar phenomenon by DRMS measurement it can be assumed that this effect is probably caused also by 'compaction' of the limestone surface by its sulphation.

7.2.4.4 Microscopic survey of limestone surface and repair mortar

A survey of the surface layers by optical and scanning electron microscopy was focused mainly on the characteristics of damage, possible remnants of paint layers, determination of crust composition and their connection to the limestone underneath as well as the structure of the limestone itself. Also the investigations of

(a)   (b)

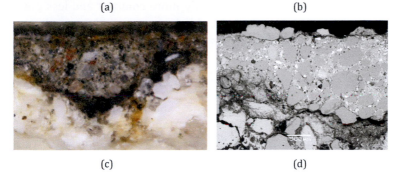

(c)   (d)

Figure 7.8 (a) Cross section, optical microscopy image, incident light at a magnification of 50 ×; black crust on limestone surface with remnants of light red and white paint layers containing red ocher and lead white. (b) Cross section, SEM-BEI; dense black gypsum crust on the limestone surface with remnants of paint layers; clearly visible detachment of the crust an paint layers from the limestone underneath. (c) Cross section, optical microscopy image, incident light at a magnification of 50 ×; layer of repair mortar on limestone surface; well visible crack between the mortar and limestone. (d) Cross section, SEM-BEI; layer of compact repair mortar on limestone surface; well visible crack between the mortar and limestone.

composition and repair mortar structure were included in this part of the assessment of the statue's condition (Fig. 7.8).

The crusts that have developed over time on the limestone surface are composed mainly from gypsum – the product of chemical limestone corrosion (the conversion of calcium carbonate to calcium sulphate). The crust is relatively compact, and according to its structure it is assumed that its properties (porosity, water absorption, water vapour permeability) are significantly different from those of the original limestone. For this reason, at the interface between limestone and crust, there is crust detachment, but also to the degradation of limestone itself. Below the crust local fragments of earlier paint layers, were found wherein the first light red layer is probably authentic. All investigated repair mortar layers applied during recent restoration work are based on Portland cement as a binder. Quartz sand was used as aggregate and the repair mortars were coloured using earth pigments in combination with carbon black. The binders are significantly more compact and less porous than the limestone. They are likely cause similar problems as mentioned in the case of crusts.

7.2.4.5 Basic stone properties related to its porosity

The measurements confirmed the expected limestone properties of high porosity and water absorption. The results of pore size distribution measurements by mercury intrusion porosimetry fits quite well with the microscopic study of the limestone cross sections (see Figs. 7.9, 7.10 and Table 7.5). The Kutná Hora limestone used in the statue is highly porous, with a relatively high proportion of large pores. The calculated average pore size is approx. 80 µm and the main part of pore sizes is in the range of 10–200 µm.

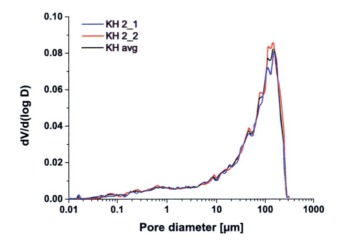

Figure 7.9 Pore size distribution (mercury intrusion porosimetry).

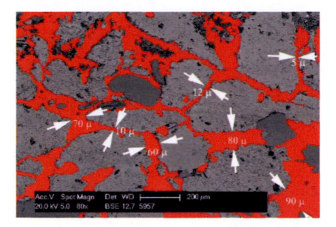

Figure 7.10 Cross section, SEM-BEI; visualisation of the porous system by red colouration.

7.2.5 Summary of the Statue Condition and Results of Scientific Investigations

In general, the investigation has shown that the statue is in a critical condition. There are irreversible changes to the surface layer and the loss of original mass. A high degree of corrosion affects the physical structure of the statue. Thick gypsum crusts are very compact and on many parts detached from the original heavily damaged stone underneath. This phenomenon has also contributed significantly to the deformation of the original shape of the statue.

The statue was originally painted. Two layers of painting remnants (light red and white) have been revealed under the crust in one of the samples.

The limestone has a high water absorption and high porosity with a relatively high proportion of large pores. The mineralogical composition is heterogeneous, the limestone is composed mainly

Table 7.5 Basic stone properties

Bulk density	1377 kg/m^3
Total water absorption by immersion	11.1 M%
Water-accessible porosity	15.3 vol%

of biodetrit (residues of mollusc shells), quartz grains, fragments of rocks and mica. The individual grains have quite weak calcium carbonate bonding and the stone is very sensitive to the acidic environment (sulphur and nitrogen oxides).

7.2.6 The Conservation Process and the Main Conservation Requirements

A conservation and restoration process for the angel statue was decided upon based on the assessment of the statue's condition. It was based also on the assumption the statue should be returned to its previous exterior location after restoration. For several reasons there was a requirement to use conservation materials which have a high degree of compatibility with the original material. Therefore an important part this process was to test and use new types of conservation materials based on calcium hydroxide nanosuspensions.

The main requirements defined for the conservation of the angel statue are:

- Stabilisation of highly damaged parts before other restoration treatments: pre-consolidation
- Removal of inappropriate repair mortar
- Reduction of gypsum crusts
- Structural consolidation of damaged areas
- Grouting, filling of cracks

The possible use of a calcium hydroxide nanosuspension was chosen for the pre-consolidation, structural consolidation and filling of cracks. Before its application, comprehensive testing was carried out on both reference limestone samples and samples simulating the highly damaged limestone.

7.2.7 Consolidation Testing of the Reference Samples

To evaluate the consolidation process two series of samples were investigated. One series was prepared directly from porous type of Kutná Hora limestone. In addition, samples simulating corroded rock were prepared by compacting of limestone sand with a grain

Figure 7.11 Test specimen samples simulating corroded limestone prepared by compacting of limestone sand.

size of 0–4 mm into moulds with dimensions of $4 \times 4 \times 4 \, cm^3$. A rather simple technique was developed to guarantee that the samples not only had the desired characteristics of high porosity and low mechanical strength, but also sufficient homogeneity which would be essential with regard to the reproducibility of the consolidation testing results (Fig. 7.11).

The selection of suitable types of nanosuspensions and optimal application methods were essential for testing on the reference samples.

The test samples were impregnated with a defined volume of CaLoSiL® E25 and CaLoSiL® IP25 nanosuspensions. These two nanosuspensions were selected as the most appropriate on the basis of a number of previous tests against other candidate nanosuspensions. Different criteria were used for the consolidation efficiency assessment. One of the fundamental criteria was the evaluation of the penetration depth of the nanosuspensions into the consolidated material. The depth of penetration was determined by colouring the fracture area of the consolidated test sample by means of 1% phenolphthalein solution in ethanol, using the alkaline reaction of the consolidant before its conversion to calcium carbonate by carbonisation (Fig. 7.12). The mechanical properties of the specimen samples, such as compressive strength and flexural strength were measured before and after the consolidation. The microstructure of consolidated samples was studied by means of optical and scanning electron microscopy. For the analysis and

Figure 7.12 Penetration depth of calcium hydroxide nanosuspension CaLoSiL® E25 determined by colouring of freshly consolidated sample fractures using 1% phenolphthalein solution in ethanol, from left to right: vertical application by dropping (2 cycles); horizontal application by dropping (2 cycles); horizontal application by brush (2 cycles); vertical application by brush (2 cycles).

examination of the microstructure, fragments of samples or polished sections were used. The measurement of the ultrasound velocity (p-waves) on model samples before and after consolidation was also performed.

7.2.7.1 Penetration depth

The penetration depth achieved by dropping the nanosuspensions onto a dry substrate was significantly greater than by application by brush. It was possible to reach a penetration depth of several cm (up to the whole volume of the test specimens) by the repetition of impregnation cycles on the used substrate by dropping. Based on the results of the penetration depth study, application by dropping was identified as the most appropriate method of applying the calcium hydroxide nanosuspension.

7.2.7.2 Carbonation rate

The carbonation rate was assessed by a gravimetric measurement of consolidated and non-consolidated reference samples (Table 7.6 and Figs. 7.13 and 7.14).

The assessment of mass gain was important for the evaluation of the final content of the consolidant and for the determination of the carbonation rate under defined conditions. The samples of a size of $4 \times 4 \times 4 \, cm^3$ were treated by dropping until their whole volume was impregnated by the nanosuspension CaLoSiL® E25.

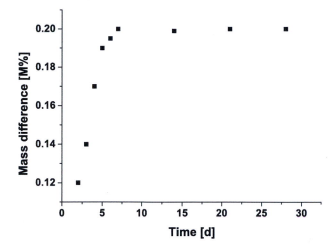

Figure 7.13 Mass gain during carbonation of consolidated samples after evaporation of ethanol.

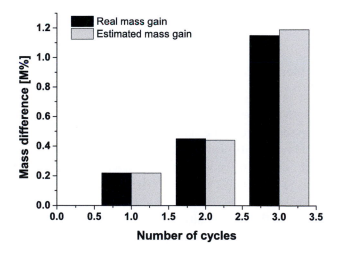

Figure 7.14 Comparison of estimated and real mass gain of the consolidated samples.

Table 7.6 Mass gain of consolidated test samples immediately after alcohol evaporation (consolidant content before carbonation)

	Limestone $(4 \times 4 \times 4\,cm^3)$		Compacted sand samples $(4 \times 4 \times 4\,cm^3)$	
	CaLoSiL®		CaLoSiL®	
Number of impregnation cycles	IP25 $\Delta m\,[\,wt\%\,]$	E25 $\Delta m\,[\,wt\%\,]$	IP25 $\Delta m\,[\,wt\%\,]$	E25 $\Delta m\,[\,wt\%\,]$
5	1.09	0.98	2.53	2.37
10	2.19	2.01	4.88	4.45

The carbonation rate at laboratory climatic conditions (T = 22°C ± 2°C, RH = 70% ± 5%) was quite high. The conversion of calcium hydroxide into calcium carbonate was completed to a high degree during 10–14 days. The real mass gain is very close to the estimated one (approx. to 90%–95%). The estimated mass gain was calculated from the volume and concentration of the nanosuspension used. There was a slight difference in the estimated and real mass gain. This can probably be attributed to the fact that there was a small amount of calcium hydroxide still present due to its incomplete conversion to calcium carbonate in the consolidated substrate.

7.2.7.3 Mechanical properties

As already mentioned above, the direct measurement of compressive strength and flexural strength was carried out in addition to the indirect assessment of mechanical properties by measurement of ultrasound velocity (p-waves). Measurements were performed after one month curing of consolidated samples at laboratory climatic conditions (T = 22°C ± 2°C, RH = 70% ± 5%).

Comparison of the results of compressive strength, tensile bending strength and the ultrasonic velocity showed a significant increase of strength of the samples after the consolidation with both tested nanomaterials (Tables 7.7 and 7.8). There is also a clear correlation between the number of impregnation cycles and the amount of consolidant, respectively (see also Table 7.6 mass gain of consolidated test samples immediately after alcohol evaporation),

Table 7.7 Strength characteristics of compacted sand samples before and after consolidation with CaLoSiL® IP25 and CaLoSiL® E25

Number of impregnation cycles	Compressive strength [MPa] CaLoSiL®		Flexural strength [MPa] CaLoSiL®	
	IP25	E25	IP25	E25
0	0.12	0.12	0.07	0.07
5	2.12	3.45	0.95	1.00
10	3.61	5.59	2.00	2.17

Table 7.8 Ultrasound velocity of limestone and compacted sand samples before and after consolidation with CaLoSiL® IP25 and CaLoSiL® E25

Number of impregnation cycles	Limestone CaLoSiL®		Compacted sand samples CaLoSiL®	
	IP25 V [km/s]	E25 V [km/s]	IP25 V [km/s]	E25 V [km/s]
0	3.12	3.12	1.13	1.13
5	3.17	3.27	2.26	2.36
10	3.24	3.49	2.39	2.70

and the increase of the strength of the samples. The results achieved with CaLoSiL® E25 are slightly better than with CaLoSiL® IP25.

7.2.7.4 Study of microstructure

The fundamental aim of the microstructure study was to assess any changes in the substrate due to the consolidation as well as to obtain an information on the consolidant's structure and its distribution.

Examination of the microstructure showed that the consolidant is deposited on the walls of the large pores and partly fills the small pores and the contact areas between the limestone particles (Fig. 7.15). This means that the consolidation does not change the overall high porosity of consolidated limestone or the relative coarse porous calcareous materials significantly and we may be expect that it will not significantly affect the properties related to water transport such as water absorption or vapour permeability.

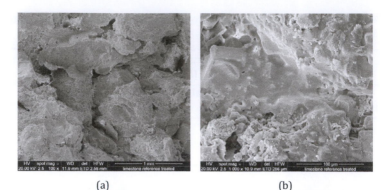

(a)   (b)

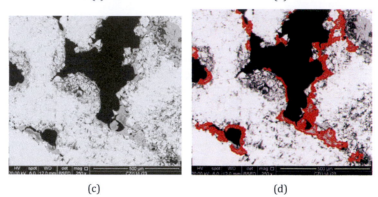

(c)   (d)

Figure 7.15 (a) SEM-BEI; fracture of consolidated limestone after curing. (b) SEM-BEI; fracture of consolidated limestone after curing, detail of a pore covered by newly developed calcium carbonate layer. (c) SEM-BEI; pore structure with newly developed calcium carbonate layer. Photo: E. Mascha. (d) SEM-BEI; pore structure with newly developed calcium carbonate layer highlighted in red.

7.2.8 Summary of Consolidation Testing on Reference Samples

The testing on both types of reference substrate has shown a satisfactory consolidant penetration depth which is the first essential requirement for effective structural consolidation. Subsequently an increase in strength has been proven and it was possible to select the appropriate type of nanosuspensions for the required purpose and, concurrently, to optimise the application method on the basis of the results. Better results were achieved with nanosuspension CaLoSiL® E25 using ethanol as a dispersal medium. This nanosuspension provided a higher consolidation rate and, at the same time, a lower tendency to form white haze on the surface of the consolidated material.

7.2.9 Conservation Treatment and Restoration of the Angel Statue

At the very beginning of the conservation treatment it was necessary to carry out a pre-consolidation, because of the substantial damage to many parts of the statue. This step was preceded by basic mechanical cleaning of the surface from the already delaminated repair mortars and ivy tendrils. Based on testing, the nanosuspension CaLoSiL® E25 was used for the pre-consolidation purposes. The most appropriate method of applying the nanolime consolidant was chosen – application by dropping using a laboratory spray bottle or syringe (Fig. 7.16). The same application method for the general structural consolidation was used after surface cleaning and partial reduction of gypsum crusts. The nanosuspension was applied according to the condition of the treated areas in 2–4 cycles up to the saturation of the material, that is, until the stone was not further absorbing the consolidant via capillary suction.

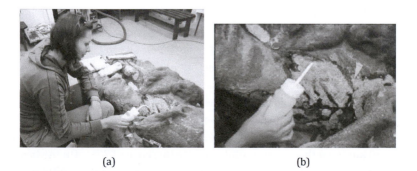

(a)   (b)

Figure 7.16 Application of the nanosuspension by dropping using a laboratory spray bottle. (a) Overview, (b) detail.

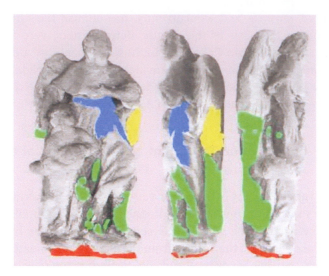

Figure 7.17 Consumption of nanosuspension for structural consolidation on different areas on the statue; yellow area: $18\,L/m^2$; blue area: $17.9\,L/m^2$; red area: $13.6\,L/m^2$; green area: not measured.

Approx. 22 L of CaLoSiL® nanosuspensions in total were used for structural consolidation of the statue of angel (height 150 cm). The consolidant consumption results for selected areas showed that the consumption ranges approximately from $14\,L/m^2$ to $18\,L/m^2$ upon the application of CaLoSiL® E25 in two cycles and four treatment cycles, respectively (Fig. 7.17).

The structural strengthening of the stone was evaluated not only visually, but also by means of several diagnostic and analytical methods. Ultrasonic velocity change (p-wave velocity) was measured by ultrasonic transmissions at selected points and the total average of ultrasonic velocity was calculated before and after consolidation (Table 7.9). As an illustration, the initial average ultrasonic velocity on the damaged statue before consolidation amounted to 1.73 km/s and the average ultrasonic velocity after the consolidation was 2.20 km/s, so there was a total velocity increase of 27%. On the basis of the assessment of the consolidant consumption and the ultrasonic velocity measurement after consolidation, it is evident that sufficient penetration depth and sufficient structural consolidation (in order

Table 7.9 Comparison of average, maximal and minimal ultrasound velocities before and after structural consolidation

Ultrasound (US) velocity	Before consolidation	After consolidation
Average [km/s]	1.73	2.20
Minimum [km/s]	0.77	1.59
Maximum [km/s]	2.75	2.72
US velocity increase - average [%]		27

of several cm) was achieved. From these observations the consolidation result can be described as satisfactory.

Particular attention was paid to the edges of damaged areas on the statue surface and the delaminated gypsum crusts. A 'modular system' of grouting was used (i.e., the use of a series of conservation materials on the same lime nanosuspension base but differing in their application or final properties, such as different consolidant concentrations) – CaLoSiL® E25 was applied as the first step, then CaLoSiL® E50 and finally CaLoSiL® paste like for filling of cracks and fissures.

After the consolidant curing, that is, the conversion of calcium hydroxide to calcium carbonate (estimated carbonation according to laboratory tests results takes 10 to 15 days), the damaged edges were stabilised by lime adhesion mortar (Fig. 7.18).

The statue surface was cleaned manually and by using micro-sandblasting, microchisel and steam cleaning after the adhesion mortar hardening (Fig. 7.19). This step included the removal of biological coverings, dust deposits, the remaining gypsum crusts and the white haze, which formed locally on less absorptive surfaces and in folds after consolidation. Based on previous experience and observation in terms of CaLoSiL® E25 application to the same type of rock under laboratory conditions, a series of tests concerning the removal of white haze was performed prior to application onto the statue. The formation of the white haze is often a negative side-effect of structural consolidation by means of calcium hydroxide nanosuspensions, especially in cases of repeated application. All tested methods – abrasive methods, wet mechanical cleaning and chemical cleaning – led to satisfactory removal of the white haze

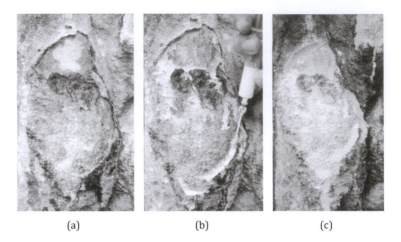

(a)   (b)   (c)

Figure 7.18 (a) Damaged part with detaching gypsum crust before grouting. (b) Injection of cracks with CaLoSiL® paste like. (c) Stabilisation of the damaged edges using lime adhesion mortar.

without damaging the limestone surface. The statue of the angel has been cleaned with a combination of wet mechanical cleaning (soft brush and controllable water vapour) and dry abrasive final cleaning (controlled micro-sandblasting). It was not necessary to use chemical cleaning methods based for example on citric acid or ammonium citrate as they pose a risk of damaging the stone surface.

Figure 7.19 Evaluation of different white haze removal methods. From left to right: 1–4: micro-sandblasting with differing air pressure and amount of abrasives; 5: manual cleaning with water and soft brush; 6: steam cleaning; 7, 8: cellulose poultice with citric acid solution (15 and 30 min. exposure).

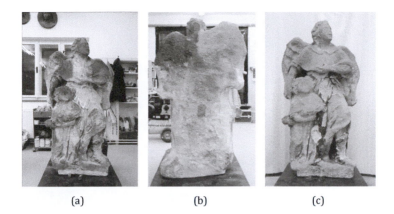

(a)   (b)   (c)

Figure 7.20 (a) Front side of the statue after consolidation with white haze formation on the surface. (b) Back side of the statue after cleaning by steam-cleaner. (c) Front side of the statue during cleaning by micro-sandblasting.

A final cleaning step followed the consolidation to minimise the mechanical stress on the corroded stone surface (Fig. 7.20).

After cleaning, the missing parts of the statue were restored with an appropriate repair mortar prepared of limestone sand and mixed binder (air lime mixed with an addition of hydraulic binders) which, in long-term proved successful on porous limestone. Finally, the statue was colour retouched where required and fillings were retouched on the surface and integrated into the surrounding colour of the stone. After the restoration and evaluation of its conservation treatment, the statue was transferred to Kutná Hora and returned to its original location (Fig. 7.21).

7.3 The Facade of the Church of the Visitation Order in Warsaw*

Restauro's adventure with nanolime began several years ago when the company joined the STONECORE project. Parallel to laboratory tests, the first CaLoSiL® pilot trials were carried out

---

*This section is written by Małgorzata Dobrzynska-Musiela.

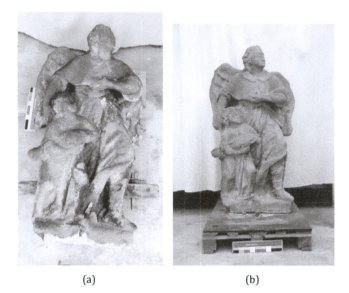

(a)   (b)

Figure 7.21   (a) Front side of the statue before restoration. (b) Front side of the statue after restoration.

on actual historic objects, including a fragment of an elevation of St. John's Cathedral in Toruń a church constructed of various materials, of which the dominant one was brick supplemented with different natural and artificial stones and with polychrome-decorated plasterwork. As can easily be inferred, the building provided a welcome opportunity to test the ability of CaLoSiL® to consolidate a whole range of materials.

At first, the experiments were conducted in small scale. They were aimed at the observation of the behaviour of the colloidal solution of calcium hydroxide applied to the consolidation of degraded areas. The preliminary results of the tests and the trials seemed promising. Soon, new opportunities arose in connection with conservation works, resulting in a large scale application of innovative formulations based on colloidal solutions of calcium hydroxide nanoparticles.

In 2009–2010 Restauro's conservation team carried out complex conservation and restoration works on the facade of the Church of the Visitation Order in Warsaw and on the western elevation of St. John's Cathedral in Toruń. These two different projects were connected by a common conservation challenge: a need to save valuable original mortars, which had survived in a state of advanced destruction. In the Toruń Cathedral, they formed the plasterwork of blind windows, while in the Visitation Church in Warsaw, they were found in the figurative and ornamental stucco decoration of the facade.

In both cases, the treatment began with an assessment of the condition of the objects, thorough investigations and development of a conservation plan. An analysis of the investigations results, together with the condition assessment and the assessment of required works, forced the conservators to seek innovative technological solutions in order to consolidate the heavily disintegrated mortars, which organosilicon preparations were not expected to have much positive impact on as would be later confirmed by field and laboratory tests. On the basis of initial consolidation tests, the colloidal solution of calcium hydroxide nanoparticles in ethanol was selected. This paper describes the related conservation treatment and explains the obtained results.

7.3.1 Introduction

The research on the application of colloidal solutions of calcium hydroxide to the preservation of historic buildings mainly addressed the problem of the consolidation of disintegrated mortar. Generally, historic mortars degrade under the influence of destructive factors faster than natural stone. Far too often, conservators decide to completely replace historic plaster or stucco decoration. This leads to irreversible loss of priceless historical fabric. For the above reasons, two elements of the stucco decoration of the facade of the church of St. Joseph of the Visitations in Warsaw (St. Augustine and the Royal Coat of Arms relief) were selected for complex restoration with the use of CaLoSiL® E25 preparation as a tool for consolidation and reinforcement of the deteriorated mortar.

The outstanding results convinced the conservators to apply this newly developed method to all the remaining stucco work of the facade. The work was carried out in 2009.

Figure 7.22 Survey of the facade with markings indicated elements consolidated with nanolime. (red) Stone elements, (blue) stucco elements.

7.3.2 Description of the Object

The church is a remarkable example of baroque sacral architecture and the only monument of the Royal Route in Warsaw that survived the devastation of World War II. Erected in two stages: in 1727–1733 (by Karol Bay or Gaetano Chiaveri, Johann Georg Plersch) and then in 1755–1761 (by Efraim Schröger or Jakub Fontana), it went through several thorough refurbishments (earlier conservation or restoration treatments were conducted in the following years: 1840–1847; 1887; 1913; 1934; 1953–1956; 1981–1983).

The facade of the church is distinguished from the other elevations by its different architectural composition and by the

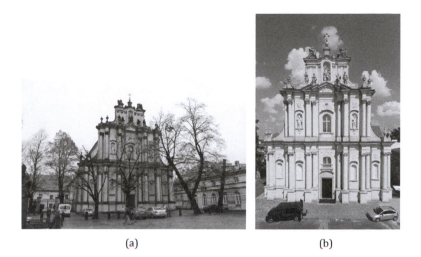

(a)   (b)

Figure 7.23 View of the facade of Visitation Order's Church (a) before conservation and (b) after conservation and restoration (2010).

division of its form. The splendid stone and stucco sculptures and decoration of the facade must have been inspired by Italian art and are related to the patron-saint of the church.

7.3.3 Technique and Technology

The facade is built of bricks laid on lime mortar. The surfaces of the walls are covered with multilayer lime plaster. The sculptured elements were chiseled in sandstone. The socles of the stories are covered with stone plates. The upper parts of the inter-story cornices and of the crowning cornice are made of stone blocks. The other decoration consists of stucco work formed with different techniques. It was drawn (cornices, bands), hand-applied, molded or cast.

The stucco work of the statue of St. Augustine was hand-applied. The binder of the mortar was lime with a little addition of gypsum, which was used to reduce contraction. The carefully selected aggregate was added to the binder in the ratio of 2 to 1. An admixture of natural glue completed the mortar. The stucco was applied on a core made of five layers of bricks. As the sculpture gained height, pieces of scorched wood were attached to it in order to minimise the total weight of the stucco work. The back of the statue is fixed to the wall of the niche which it is placed in.

The stucco work reliefs were made of lime-sand mortar, the composition of which was similar to that of the statue. The stucco was hand-applied directly on the wall layer after layer. The decoration was hand-formed. Charcoal was added in the case of higher reliefs. Probably neither the statues nor the reliefs were originally paint-coated. The surface was only polished and coated with milk made of the same mortar binder as the other stucco work. The addition of natural glue might have broken the white of the mortar and warmed its colour.

7.3.4 Program of the Pre-Treatment Investigation: Results of Petrographic Analysis of the Mortars

The aim of conservation research was to determine the monument's condition and the causes of its deterioration, as well as to identify its structure and the materials it was built of. It was also necessary to examine the colouring of the facade. The following factors were examined:

- Dampness and salinity of the building materials
- Composition of the deposits
- Composition of the mortars (chemical tests and petrographic analysis)
- Kinds of stone used for the decoration
- Stratigraphy of the mortars and the paint coats

Two groups of stucco work mortars were defined: mortars with a pure carbonate binder and mortars with a carbonate-gypsum binder. A petrographic analysis was conducted on eleven samples of mortars from the Church of St. Joseph of the Visitationist in Warsaw. The analysed samples differ mainly in the mineralogical composition of the binders and in the number of grains in the aggregates (framework grains). The composition of the aggregates is very similar. It is dominated by detrital grains, among which quartz is most numerous. Feldspars and lithic fragments are also present, and so are few accessory constituents such as grains of opaque minerals. Morphologically, the grains of the aggregates differ only a little. Their size is similar, and they show roughly the same roundness degree, being rounded or subrounded. On the other hand, the composition of the binders is mixed. They consist of gypsum and calcium carbonate microcrystals in the form of micrite. According to the ratio of gypsum and micrite, the samples can be divided into two main subgroups. In group one a gypsum binder is dominant, while the amount of the aggregate is small. In group two, the carbonates prevail, while the level of aggregate grains is relatively high.

A sample of definitely original, that is, dated c. 1750, stucco mortar was taken from the St. Augustine's statue. The mortar was classified as group 2 as its binder proved to be based on lime. The mortar composition is presented below (Fig. 7.24).

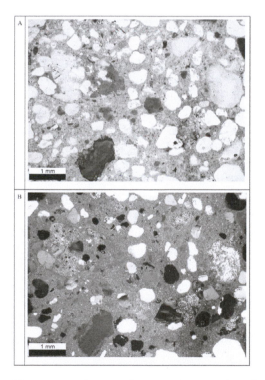

Figure 7.24 Microscopic image of a mortar sample taken from St. Augustine's statue as observed with one polariser (a) or two crossed polarisers (b).

7.3.4.1 St. Augustine's statue

Original mortar sample date ca. 1750. Description: Whitish grey mortar sample, bound, with dispersed framework grains. The largest grains do not exceed 1.0 mm and show a good degree of roundness – rounded or subrounded. The binder is mixed and heterogeneous, consisting of sub-microscopic crystals, which form numerous monocrystallic gypsum and micrite (Fig. 7.24).
Sample composition:

- Binder 67%
- Quartz 25%
- Feldspars 0%
- Lithic fragments 3%
- Other less than 0.5%
- Pores 3%

The sample taken from the decoration fragment showing the Royal Coat of Arms (John II Casimir Vasa and Marie Louise Gonzaga) differ from each other significantly and represent groups 1 and 2. The mortar from the background, probably original, is based on lime, whereas the mortar taken from a fragment of the crown contains gypsum binder. The composition of both mortars is presented below.

Group 1: Original background mortar sample dated ca. 1750. Description: Whitish grey mortar sample, weakly bound with a chaotic structure and dispersed framework grains. The largest grains measure ca. 1.0–1.2 mm. The grains show a considerable degree of roundness – they are rounded and subrounded. The microcrystallic binder mainly consists of calcium carbonate present in the form of micrite. Quite significant local cracking, fine cracks continue between neighbouring framework grains. Monomineralic fragments can be found, consisting of fine gypsum scales, which clearly contrast with the surrounding micrite (Figs. 7.25 and 7.26).
Sample composition:

- Binder 42%
- Quartz 49%
- Feldspars 1.5%
- Other 1%
- Pores 5%

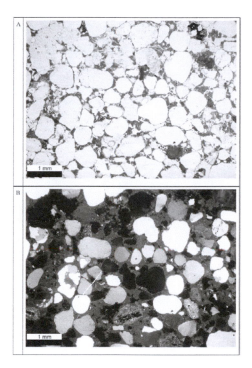

Figure 7.25   Microscopic image of a sample taken from the background of the Royal Coat of Arms as observed with one polariser (a) or two crossed polarisers (b).

Group 2: Crown mortar sample. Description: Whitish grey mortar sample, bound, with a chaotic structure. Highly dispersed framework grains. With the exception of gypsum and micrite concentrations, detrital components do not exceed 0.4 mm. Detrital components considerably rounded – grains rounded and or subrounded. Binder consisting of sub-microscopic gypsum crystals forming a heterogeneous mass with dispersed carbonates which constitute monocrystallic micrite concentrations.
　　Sample composition:

- Binder 98%
- Quartz 0.5%
- Feldspars 0.5%
- Other 0%
- Pores 1%

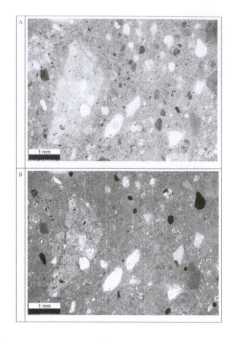

Figure 7.26 Microscopic image of a sample taken from the crown of the Royal Coat of Arms as observed with one polariser (a) or two crossed polarisers (b).

7.3.5 Condition of the Stucco Work

The stucco decoration, consisting of full sculptures as well as floral and heraldic ornaments, was found in a very bad condition.

The lime-gypsum and gypsum stucco work underwent deterioration processes much faster than the stone elements. For this reason it was often largely restored and recompleted. Different kinds of mortars were used for this purpose. During the refurbishment conducted under Enrico Marconi's supervision, lime, lime-gypsum and purely gypsum mortars were applied. However, in the later years very strong and tight mortars, the same as for refurbishing the plaster, were used, which degraded the condition of the plaster. In the 1980's, cement mortar was used to reconstruct entire compositions or large elements, whereas epoxy putties and fillers were used to fill minor losses.

The causes of such a poor condition of the stucco work are complex. Yet, the main cause seems to be improper conservation – using strong and tight mortars for filling and finishing off the surfaces. The cement mortars laid on a weaker and unstable base, that is, lime-gypsum mortars, formed a shield, in which cracks and scratches began to appear soon due to its completely different physical properties. While these cracks and scratches let water penetrate to the inside, evaporation of the water was hindered by the tight surface. The binder of the original mortars diluted and the process of freezing fractured the mortars. Stabilising the movable elements of the stucco work decoration with steel rods turned out to be very invasive due to corrosion. The rods were already installed during the refurbishment of 1847 and later in 1981 when they were plastered with very strong cement or cement-resin mortars (Figs. 7.27 and 7.28).

The plaster cast of the stucco work was disintegrated and exfoliated. There were voids in its structure. Cracks and splits occurred in places where materials of different technological composition and of different physical properties were joined. Whole fragments of the stucco work were unstable. Again, the main cause such a bad condition seemed to be improper conservation. Stabilisation of the disintegrated ground was the key treatment to apply. It success determined the possibility to carry out any further conservation.

7.3.6 Plan of Conservation Treatment

The conservation work was aimed at:

- Eliminating the causes of deterioration
- Re-establishing the functional and utilitarian properties of individual elements
- Preventing any further deterioration
- Re-establishing the esthetic and artistic value of the monument

A plan of conservation treatment was developed that was based on the concept of applying purely lime materials to reinforce and consolidate the stucco mortars. It prescribed the performance of the following operations:

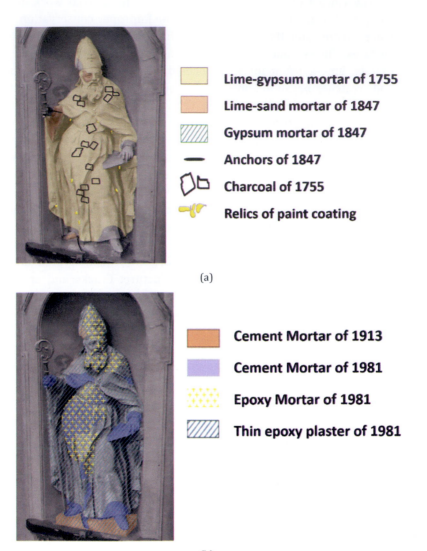

(a)

(b)

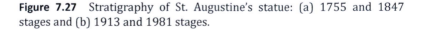

Figure 7.27 Stratigraphy of St. Augustine's statue: (a) 1755 and 1847 stages and (b) 1913 and 1981 stages.

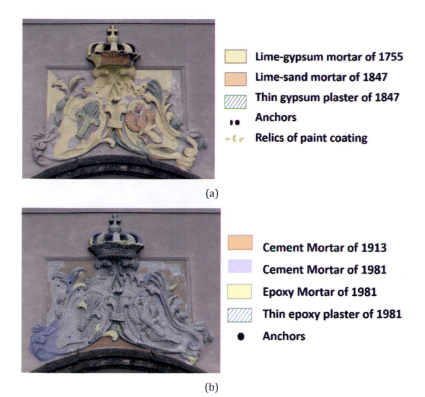

(a)

Lime-gypsum mortar of 1755
Lime-sand mortar of 1847
Thin gypsum plaster of 1847
Anchors
Relics of paint coating

(b)

Cement Mortar of 1913
Cement Mortar of 1981
Epoxy Mortar of 1981
Thin epoxy plaster of 1981
Anchors

Figure 7.28 Stratigraphy of the Royal Coat of Arms: (a) 1755 and 1847 stages and (b) 1913 and 1981 stages.

- Mechanical cleaning of the paint coating and spackling pastes from the stucco work surfaces
- Mechanical removal of the technologically faulty cement and epoxy-based fillings; removal of the corroded anchors and frames; simultaneous, preliminary reinforcement and gluing of the degraded original fabric
- Dismantling of loose elements of the stucco decoration
- Structural reinforcement and consolidation of the weakened parts of the stucco work
- Stabilisation and gluing of the movable fragments of the stucco work

(a)   (b)

Figure 7.29 St. Augustine's statue (a) before and (b) after conservation and restoration.

- Filling the cavities and reconstruction of the missing forms
- Final treatment of the surfaces
- Painting of the stucco work surfaces

7.3.7 Conservation Treatment

After the removal of deposits and fillings, the condition of the early forms dating back to 1755 and 1847 was found to be very poor. Light mortars based on lime and gypsum binders required preliminary consolidation and stabilisation before they were able to undergo any further treatment. Reinforcement was performed with the use of CaLoSiL® E25.

Different methods of application were employed. The preparation was brushed in, poured over the degraded structure of the mortar, injected and applied with a system of drips.

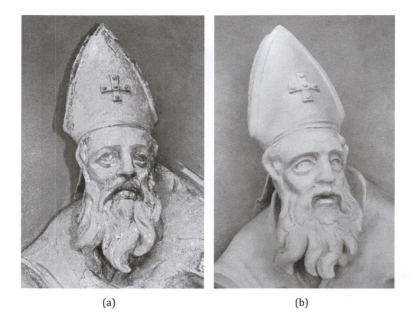

(a)   (b)

Figure 7.30 (a) St. Augustine's head before conservation, visible are the losses and scaling of the superficial stucco layer and paint coating, cracks and fillings. (b) St. Augustine's head after conservation and restoration.

The sculpture elements which had been taken down during the cleaning were saturated with CaLoSiL® through immersion. Although all methods of application proved to be effective, dripping surpassed the other ones. The preparations penetrated disintegrated mortars very well. About 20 L of CaLoSiL® E25 preparation was introduced into the statue of St. Augustine and 5 L into the Royal Coat of Arms. An atmosphere of increased humidity, necessary for the proper binding of the calcium-based preparation, was obtained through sprinkling the surface of the stucco work elements with water. The procedure was frequently repeated, while the condition of the mortar being reinforced was constantly monitored.

At the same time, during the impregnation, loose, friable, and exfoliated elements of the stucco work were glued with two preparations used interchangeably: an injection mass based on dispersed

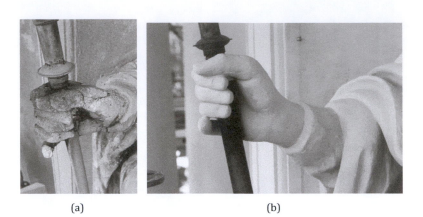

(a)   (b)

Figure 7.31 (a) St. Augustine's palm before conservation – mainly not original, reconstructed with a mortar with an addition of cement, superficial layers showing microkarst and stained with compounds coming from the corroding copper pastoral staff; largely cracked and unstable. (b) St. Augustine's palm after conservation and restoration.

(a)   (b)

Figure 7.32 (a) The Royal Coat of Arms of John II Casimir Vasa and Marie Louise Gonzaga before conservation – the surface is washed out, paint and finishing coatings are scaling; losses and filling are visible; deposits accumulated in high-relief fragments. (b) The Royal Coat of Arms after conservation and restoration.

Figure 7.33 A fragment of the Royal Coat of Arms crown, visible is the strong scaling of the paint and finishing coating and minor losses; the surface soiled significantly.

Figure 7.34 A fragment of the Royal Coat of Arms before conservation, visible are the strong scaling of the paint and finishing coating, losses and erosion; numerous fillings, soiled surface.

Figure 7.35 A fragment of the Royal Coat of Arms before conservation, visible are the strong scaling of the paint and finishing coating, losses and erosion; numerous fillings, soiled surface.

Figure 7.36 St. Augustine's head during treatment – after the removal of technologically and technically faulty fillings, extensive material losses and the weakening, disintegration and crumbling of the mortar became evident.

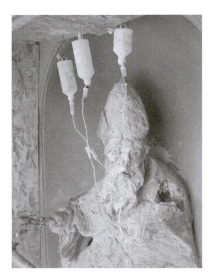

Figure 7.37 Structural impregnation with CaLoSiL® E25 introduced through a drip into the stucco mortar.

Figure 7.38 The Royal Coat of Arms during treatment after the removal of numerous secondary, faulty fillings; visible are cracks, delamination, splitting and the strong disintegration of the mortar.

Figure 7.39 The Coat of Arms during reinforcement with CaLoSiL® E25 through a drip.

Large reconstructions.

Saturation with a reinforcing preparation injected through splits and exfoliations.

Saturation with a reinforcing preparation applied with drips.

Anchors

Figure 7.40 A map of conservation treatments performed on St. Augustine's statue.

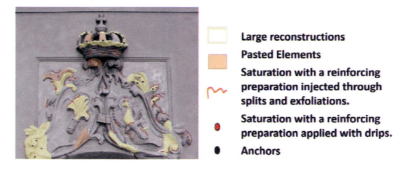

Large reconstructions

Pasted Elements

Saturation with a reinforcing preparation injected through splits and exfoliations.

Saturation with a reinforcing preparation applied with drips.

Anchors

Figure 7.41 A map of conservation treatments performed on the Royal Coat of Arms.

calcium hydroxide, containing properly selected modifiers and a second mortar in the form of powder, which was based on natural lime and contained no soluble salts, but only an addition of properly selected fillers and modifiers with rheological properties.

For filling the losses, a special mortar was prepared on the spot. It was purposely composed to correspond to the properties of the original fabric. The mortar was based on three-year aged pitted lime (crema di calce). The lime was mixed with fine river sand (grains up to 0.8 mm) in the ratio of 1 : 3.

Due to the number of the historical and modern fillings, it was necessary to unify the surface of the stucco details and sculptures. The surface was laid with a putty based on lime mortar (containing historical lime filler fine combined with carefully selected river sand, whose grains measured between 0.5 and 1.2 mm). Then the stucco elements were coloured in the same way as the original stucco mortar. It was laid in such a manner as to resemble the original surface as much as possible.

7.3.8 Assessment of Treatment Results

Overall consumption of CaLoSiL® E25 for both elements was 25 L, which proved the capability of the preparation to penetrate deep into the structure of impregnated mortars. St. Augustine's statue with the surface measuring 7.76 m^2 required 20 L of CaLoSiL® E25, while the Royal Coat of Arms, whose surfaces measures 3.02 m^2 – 5 L.

In order to examine the degree of the preparation penetration into the structure of the mortars, pre- and post-treatment samples were collected for an SEM-BSE analysis. A comparative microscopy study was conducted by Elisabeth Mascha (Universität für angewandte Kunst Wien, Austria). The microscopy study (SEM-BSE analysis) confirmed deep penetration of the solutions of colloidal calcium hydroxide nanoparticles into the structure of the mortars (Figs. 7.42–7.47).

The performance of an impregnation procedure simultaneously with consolidation and gluing was possible owing to the preparation chosen for reinforcement, namely CaLoSiL® E25 and its colloidal calcium hydroxide base. The affinity of the binders, the impregnate and the injection masses allowed efficient work without technological pauses (required when preparations based on silicic acid esters are implemented in the impregnation process) as well as safe continuation of conservation treatment.

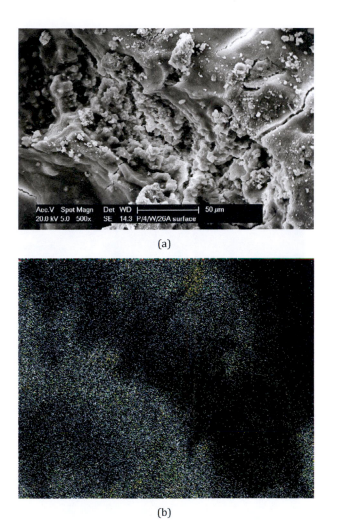

(a)

(b)

Figure 7.42 SEM micrographs of a nanolime consolidated stucco mortar sample taken from the Royal Coat of Arms. Photos: E. Mascha.

The choice of nanoparticle calcium hydroxide in a colloidal solution allowed simultaneous work with a system of lime-based materials for reinforcing, consolidating and filling stucco work.

Post-treatment investigation of the mortars proved the effectiveness of CaLoSiL® E25 preparation in consolidating friable mortars.

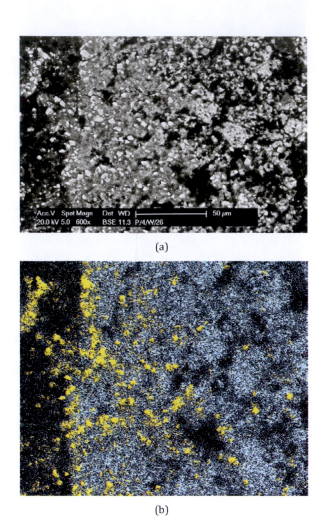

(a)

(b)

Figure 7.43 SEM micrographs of a nanolime consolidated stucco mortar sample taken from the Royal Coat of Arms. Photos: E. Mascha.

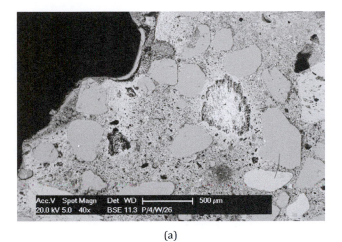

(a)

(b)

Figure 7.44 SEM micrographs of thin sections of nanolime consolidated stucco mortar samples taken from the Royal Coat of Arms. Photos: E. Mascha.

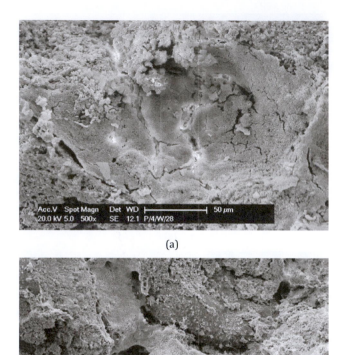

(a)

(b)

Figure 7.45 SEM micrographs of a nanolime consolidated stucco mortar sample taken from the Royal Coat of Arms. Photos: E. Mascha.

Figure 7.46   St. Augustine after conservation and restoration.

Figure 7.47   The Royal Coat of Arms after conservation and restoration.

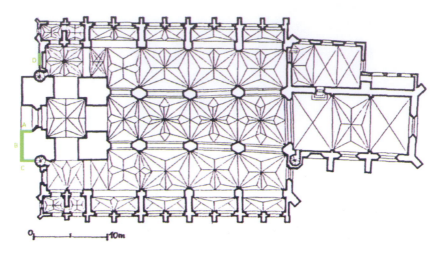

Figure 7.48 Section of the cathedral with markings indicating fragments nanolime was applied to during tests and treatment. The letter D area includes the blind window which underwent treatment.

7.4 The Western Elevation of St. John's Cathedral in Toruń*

7.4.1 Introduction

During the STONECORE project, the reference object – the Tower of St. John's Cathedral (western wall of the northern extension) was repeatedly examined and tested. After the analysis, a blind window covered with plaster was selected for demonstration. Although the chosen fragment was not large ($13.65\,\text{m}^2$), treatment was divided into two stages (Fig. 7.48).

The first one, carried out in July 2010, was focused on conducting preliminary examination and consolidating a few test areas with nanolime, SAE and both preparations simultaneously. In September the same year the areas which nanolime was applied, were examined (also in situ). In summer the following year, the research was continued in order to determine the optimum impregnation methods and conditions. Then the whole blind window was comprehensively treated.

---

*This section is written by Małgorzata Dobrzynska-Musiela.

(a)   (b)

Figure 7.49   (a) A fragment of the northern wall of the tower annex (area D) with the blind window before treatment. (b) Orthophotoplan of the blind window before treatment.

Figure 7.50 A general view of the Toruń St. John's Cathedral from North-West after conservation and restoration.

7.4.2 Object Description

The Cathedral Basilica of St. John the Baptist and St. John the Evangelist in Toruń, the old parish church of Toruń's Old Town, provided space where local authorities were elected, patricians were buried and the most important celebrations were held, including those connected with royal visits. Till the end of the 20th century, it was traditionally referred to as St. John's Church, in 1935 it was given the title of minor basilica, and since 1992 it has been the Cathedral of the Toruń Diocese. The church was built in a few stages, the earliest one dating back to the second half of the 13th century. The final structure consists of a low presbytery, which contrasts the massive three-nave body with lower chapels and the enormous western tower. The king John I Albert's heart is buried in the church. Also Nicolaus Copernicus is commemorated there – one can find his baptismal font, an epitaph dedicated to him and an 18th century statue of the astronomer. In the Tower hangs the third largest bell in Poland, Tuba Dei, which was cast in 1500.

The church is one of the finest examples of gothic architecture found in the Chelmno Land. It consist of 27.30 m tall three-nave hall with a lower presbytery and adjacent chapels. A massive quadrilateral tower blends with the church's main body in its western part. The tower's western elevation is marked with a deep niche going from its base up to the roof. Extensions with church porches flank the tower. They are lower than the naves. The presbytery has stellar and groin ribbed vaults, the nave body enriched stellar vaults (in the nave, 8-pointed stars; in the aisles, 5-pointed ones).

7.4.3 Technique and Technology

The foundations of the building are made of stone, walls of solid, fired brick laid on lime mortar in the Gothic and Monk bond. The plaster on the facade blind windows (bays and friezes) was originally decorated with architectural polychrome.

The blind window, originally open, was bricked in 15th century and then covered with a thin layer of whitewash and plastered. The plaster was painted red.

7.4.4 Condition

The conditions of individual elements made of different materials (shaped brick framing the blind window, brick wall, lime mortar, and plaster) varied. The brick wall was cracked and stratified due to a construction collapse in the past and previous reconstruction efforts. The shaped bricks framing the blind window were covered with greyish black and black crusts of atmospheric origin and of different thickness and degree of adhesion to the ground.

Plasters survived in a rather poor condition. They were covered with black deposits and they lost adhesion to the ground. In some areas they were falling off; there were cracks and minor cavities. Layers of colour (red) were observed on the surface.

The wall in which the blind window is located, suffered serious structural damage probably due to two construction disasters that had caused the tower to collapse. A structural break, which was partially shifted, ran through most of the height of the wall and penetrated throughout its thickness. The break reached the finial of the pointed-arch blind window, which caused damage and losses of

Figure 7.51 Orthophotoplan of the blind window with graphical and colour markings indicating building materials. (blue) Sandstone, (yellow) gothic brick, (grey) lime plaster, (dark grey) cement mortar.

the shaped bricks framing the blind window. The break continued along the left edge of the blind window. Beside the damage which had resulted from the movements of the structure, the brick bed and the framing were in a relatively good condition. Single bricks and shaped bricks (misfired) were weakened. The red painting layer,

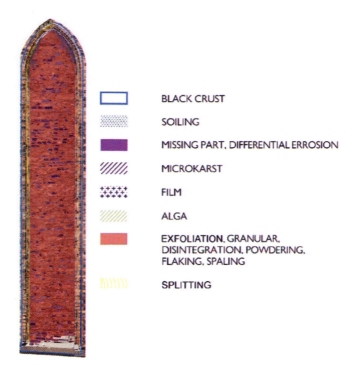

☐	BLACK CRUST
▦	SOILING
▮	MISSING PART, DIFFERENTIAL ERROSION
▨	MICROKARST
▨	FILM
▨	ALGA
▬	EXFOLIATION, GRANULAR, DISINTEGRATION, POWDERING, FLAKING, SPALING
⁝⁝⁝	SPLITTING

Figure 7.52   Orthophotoplan of the blind window with graphical and colour markings indicating degradation symptoms.

which covered the plaster, had been completely washed out. It had been preserved only in crevices and hollows in the plaster. The face of the plaster had also been washed out. The surface of the plaster was covered with dirt. In some parts atmosphere-originated deposits formed a hard, black, sealing crust. The plaster was cracked and partially detached from the bed. Under its face the weakened plaster crumbled and powdered.

7.4.5 Program of the Pre- and Post-Treatment Investigation

The investigation program was developed in collaboration with ITAM (Institute of Theoretical and Applied Mechanics of the Czech

Figure 7.53 A fragment of the blind window showing degradation phenomena.

Figure 7.54 A fragment of the blind window showing degradation phenomena.

Figure 7.55 A fragment of the blind window showing degradation phenomena.

Academy of Sciences of the Czech Academy of Sciences, Prague) experts, who carried out most of the works. The study encompassed the identification of primary and secondary materials found in the blind window as well as the establishment of their basic physical properties. The plasterwork was examined in terms of:

- Composition and structure (petrographic, microscopic, thermo-gravimetric analysis)
- Ethanol and CaLoSiL® absorption (Karsten tube)
- Porosity (porosimetric analysis before and after impregnation – mercury porosimeter)
- Mechanical properties before and after impregnation – strength in bending (flexural strength)
- Reinforcement effect (drilling resistance test)
- Surface cohesion after consolidation test (peeling test)

A series of tests was also carried out in order to determine the optimum CaLoSiL® application method. The research was focused on reducing the white lime haze phenomenon and on finding the most effective method of removing the haze, should it occur during treatment.

7.4.6 Investigation Results

7.4.6.1 Identification of the materials

The research showed that the blind window ledge was made of quartz arenite. It features are described below. Description: 15th century blind, window ledge; sandstone-quartz arenite, colour is grey, sample is compact, structure of the rock is psammite, even-grain structure, mineral composition consists of quartz, tourmaline, zirconium, rutile; the framework grains are mainly quartz, size of grains does not exceed 1.0 mm, grains are mainly sized 0.1–0.4 mm. Morphology of grains: slightly elongated or isometric, roundness rate is good, grains are usually subrounded to sub-angular, type of binder is regenerative, contact type; siliceous.

Sample composition:

- Quartz 78.5%
- Feldspars <0.5%
- Other 3%
- Binder pores 18%

The dimensions of the bricks that fill the blind window are of vary ($28/29 \times 12/14 \times 7.5/9$ cm) and so is their degree of firing and the material used to produce them. The colour of the solid bricks and the shaped bricks differs from orange to dark red.

The mortar – the most interesting material from our point of view – that was used to plaster the blind window was based on hydraulic lime, what was determined by thermogravimetric analysis (TGA).

The fillers consisted mainly of: quartz medium, well-selected, dipped differently, feldspars, dark unidentified minerals, fragments of ceramics, fibre, wood, vegetable fibres. The granulation was determined between 0.02–0.1 mm. The water absorption of the mortar had an average of 10.5%. The microstructure of the plaster mortar was examined in detail at ITAM. The following image

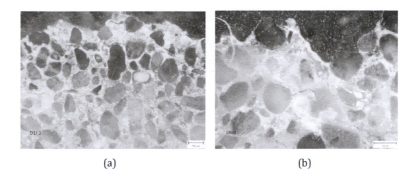

(a)   (b)

Figure 7.56   Thin sections of plaster.

captures lime render (D1, D5, D6). The aggregate content was 70 wt%, mainly based on fine river sand, grain size <0.2 mm and quartz as the main mineral. The homogeneous binder was determined as 30 wt%. The following surface characteristics were made: grey-black layer of thickness 1.0–2 mm; white particles (gypsum) in the binder; in a layer 0–0.5 mm under the surface cracks of 30–100 μm thick. Detail of cracks $d = 31$ μm sufficient for penetration of CaLoSiL$^®$; cracks in surface zone 0–0.5 mm.

Surface zone 0–0.5 mm enriched with gypsum (white particles in the binder). Likely signs of atmospheric pollution. Outer layer can be decisive for water uptake properties.

7.4.6.2 Absorbability test of plasters with ethanol and CaLoSiL$^®$ E25

Test results showed that capillary rise rate of both liquids applied by means of Karsten tube was relatively long. The rise rate for nanolime was however much longer than for ethanol. This may partly be due to certain tightness of the plaster surface resulting from atmospheric deposits (gypsum layer). The results are presented in Table 7.10.

The above observation and the analysis of the condition of the plaster led to the decision that at the next treatment stage the preparation should be injected through existing cracks and fissures, which is always an effective application method.

Table 7.10   Absorbability test of plasters with ethanol and CaLoSiL® E25

Sample index	Capillary rise rate Ethanol	Capillary rise rate CaLoSiL® E25
D/9	1.0 cm–13'	0.5 cm–40'
	2.0 cm–24'	
D/ Cal	3.0 cm–39'	
	4.0 cm–46'	
D/10	1.0 cm–7'	0.5 cm–45'
	2.0 cm–15'	
D/10 Cal	3.0 cm–25'	
	4.0 cm–33'	
D/12	1.0 cm–5'	0.5 cm–23'
	2.0 cm–10'	
D/12 Cal	3.0 cm–16'	1.0 cm–55'
	4.0 cm–23'	

7.4.6.3 Porosimetry

A key property of renders is their porosity as the interconnectedness and thickness of pores determine the inflow of liquids carrying the consolidant. The porosity was measured by means of a mercury porosimeter. However, in order to get representative evaluation of the whole system behaviour, that consists of the surface contribution and bulk pore properties summarised in one number, it is preferable to use water uptake properties measurable, for example, by the so-called Karsten tube. On the other hand using porosity information from the different depths one can learn valuable data on differences between bulk and surface layers and the difference between untreated and treated surfaces. Figure 7.57 on the right margin of the page show locations from which specimens were acquired.

Two sets of specimens are clearly distinguishable from the image marked by grey and red numbers. The grey ones represent untreated mortar, while the red one was treated by CaLoSiL®. When the two sets (the grey one also averaged) are compared, on the plot, the two pore distributions are placed one above the other, it can be inferred from the plot that the main effect of CaLoSiL® treatment can be seen in filling of small pores. Manifestation of this process can be seen on red plot which does not contain any pores below

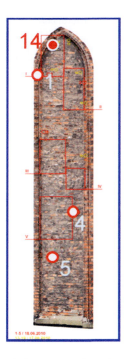

Figure 7.57   Areas of plaster sampling for porosimetric tests.

3 μm diameter. Implications of this finding are twofold: surface consistency of the treated area was improved, but, on the other hand, another treatment is unlikely to penetrate into the mortar, while leaving a white haze on the surface.

Table 7.11 summarises relative pores distributions in the material removed from various locations on the cathedral's blind window, shows clearly effect of consolidation. The CaLoSiL® has

Table 7.11   Porosimetry

	Division of pores					
Diameter	0.001–0.01	0.01–0.1	0.1–1	1–10	10–100	100–1500
Toruń_D1	3	10	20	34	16	18
Toruń_D4	2	5	28	42	11	12
Toruń_D6	4	6	21	39	16	13
Toruń_D14	0	0.1	0.2	20.3	31.4	48.1

Table 7.12 Characterisation of the samples

Specimen	Density[g/cm^3]	Open porosity [%]
Toruń_D1	1.81	32
Toruń_D4	1.81	31
Toruń_D6	1.79	34
Toruń_D14-surf	2	16

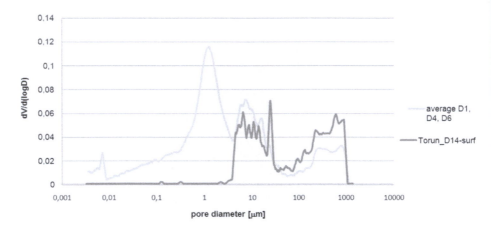

Figure 7.58 Pores distribution in mortar specimens in the plaster samples.

closed some kind of pores – those of micron and sub-micron sizes and also reduced the overall porosity of material.

The second column from the right side of table is the most important for CaLoSiL® penetration: it shows that the total porosity has been halved.

7.4.6.4 Strength in bending flexular strength

Extracted render specimens were extended by wood prostheses in order to form sufficient base for three-point bending (3PB) tests (see Fig. 7.59). This simple method enabled to measure 3PB strength of mortars and renderings, that would be otherwise impossible to study, due to limited specimen dimensions implied by small thickness of the rendering itself. The following Table 7.13 presents 3PB test results for specimens extracted from the same

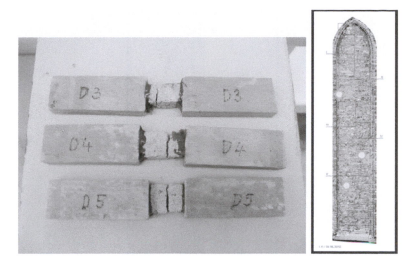

Figure 7.59 Areas of sampling for strength tests and an after-fracturing image showing the method of sample preparation for tests.

mortar but subjected to different treatments by consolidants. The results demonstrate that treatment effects and variation in material strength are hard to separate.

The strength itself exhibits stochastic behaviour and therefore Weibull's probabilistic description would be appropriate with accordingly increased sample = number of specimens in set. Considering the fact that 3PB is not a non destructive method, complying with the requirements of statistics one would fail conservation criteria not to cause any unnecessary harm to the object of restoration. That is why strength data play only marginal role in this study, especially if the main target is improvement of surface mechanical properties.

Table 7.13 Characterisation of the samples

Sample	Treatment	Flexural strength
D3	2x E25	1.8 MPa
D4	2x E25 + KSE	1.5 MPa
D5	2x KSE	2.7 MPa

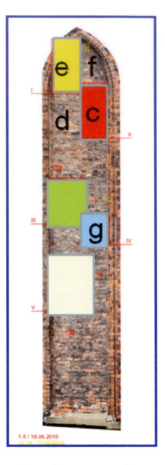

Figure 7.60 Areas where plaster was peeling-tested for cohesion.

7.4.6.5 Peeling test summary

Using such a simple technique as peeling test, one can obtain illustrative results distinguishing quality of studied surface even for inexperienced observer. As the sensitive laboratory scales is the only requirement of the method, it seems that it can be successfully used in conservation areas.

The plot (Fig. 7.61) summarises collected data and supports above mentioned claims for possibility of effortless determination of surface state by the plot curvature. Aggregate plot clearly depicts

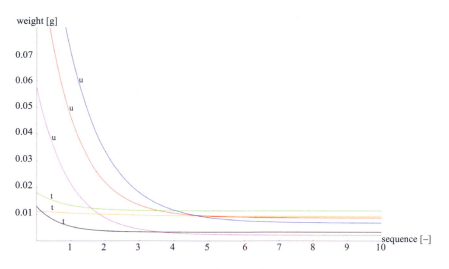

Figure 7.61 Graph showing results for the surface before and after consolidation.

differences in behaviour between treated (marked *t*) and untreated surfaces (marked *u*).

Legend A: magenta, B: orange, D: red, E: green, F: blue, G: black. Peeling tests A and B were carried out on the test area A (not described in the section), while the tests D through G were done on the area D depicted on the image above.

7.4.6.6 Drilling resistance summary

It can be observed that the influence of internal inhomogeneities of the material to drilling resistance is more important than the effect of consolidants itself. For this reason, it is necessary to accompany every parameter with an empirical interpretation to assess the specific case of measurement. Another implication of this situation is that averaging several tests would not lead to meaningful values (statistics would yield a useful data for at least a several dozen of drillings performed in one surface).

Overall, when applicable, the impact of consolidants is particularly evident in the near surface layer (0–3 mm) in depth. Frequently is this effect obscured by random distribution of inhomogeneities in the test samples, and therefore one value representing overall material's performance cannot be reliably established.

The available literature still lacks a description of a methodology that would eliminate the influence of inhomogeneities. One possible method would be eliminating of the peaks by envelope a curve connecting with some smoothness the minima in drilling resistance plot and declaring that this curve is lower bound of (softer) matrix (binder) approximation.

As concerns area D offers more pronounced effect of the consolidation, but only after careful comparison of drilling resistance plots. Some surface improvement of probes in treated areas is detectable in terms of increased drilling resistance close to surface.

7.4.6.7 Investigation into optimisation of the application method

To determine the most effective application method, subsequent experiments were conducted on the plastered blind window in the areas impregnated in July 2010. Part of the plaster was saturated with CaLoSiL® E25. The procedure was effective in terms of strength increase. Unfortunately, the visual effect of those tests could not be accepted owing to haze which appeared on the surface of the saturated plaster. Therefore the following aims were set for the next series of tests in July 2011:

- Determining a method of removing the haze
- Determining an application method which would reduce the occurrence of haze

Three methods of removing haze were tested:

- Dry mechanical or abrasive method: It used properly selected aggregates applied under controlled pressure with a micro-sandblaster
- Wet mechanical method: It used a jet of water vapour applied under pressure with a water vapour generator
- Chemical method: It used poultices of 5% citric acid

Figure 7.62 The result of removing white haze from the plaster surface with an abrasive blasting method.

The abrasive method was definitely the most effective (Fig. 7.62). Almost all the white deposits were removed without damaging the surface of the plaster. The other methods were not effective.

The second stage of the research concerned the development of an application method that would eliminate or reduce the formation of haze on impregnated surface.

The experiments were based on the methods established by our STONECORE partners, which are detailed summarised in various chapters of this book. The method consisted in lowering the preparation concentration by half (12.5 g/L) and adding a little hydroxypropyl cellulose to the preparation; alternatively an HPC solution (0.2–0.5% in ethanol) could be applied to the impregnated surface right after impregnation. The aim of the experiment was to verify the repeatability of the results obtained by the specialists from Dresden. The obtained results were very similar. Diluting CaLoSiL® E25 in ethanol to 12.5% reduced the occurrence of haze considerably. If the surface of the plaster was treated with 0.5% HPC solution after each application, there was almost no haze on the impregnated surface. However, it is important to examine whether

Figure 7.63 White haze after CaLoSiL® E25.

this desirable visual effect is not due to the clogging of the surface pores of the impregnated material, which limits evaporation and the possible migration of calcium hydroxide to the surface. There exists a possibility that the hydroxide accumulates right under the surface of the impregnated material.

7.4.7 Plan of Conservation Treatment

After the tests and experiments were completed, conservation treatment of the blind window began. Guidelines, treatment plan, and conservation aims were developed first. The blind window needed to be structurally reinforced. The structural breaks in the wall had to be filled. Optionally, they could be tied with steel-resin anchors. Loose shaped bricks had to be be resettled. Any missing bricks were to be replaced. The plaster required reinforcement and consolidation. After the completion of technical works, missing plasters were replaced and the surface was consolidated and

Figure 7.64 Plaster surface after being impregnated twice with colloidal calcium hydroxide diluted in ethanol in the proportion of 12.5 g/L and subsequently treated with HPC in ethanol (0.2%).

secured with a lime whitewash of the same colour as the old plaster. The following treatment and conservation program was realised:

(1) Removing the loosened fragments of the shaped bricks
(2) Preparing cracks and crevices for consolidation by removing all the disintegrated material and dirt and by following disinfection
(3) Injecting crevices and hollows in the exfoliated wall with a mineral mortar
(4) Optional tying of the breaks with steel-resin anchors
(5) Repairing the finial of the pointed-arch blind window – replacing any missing shaped bricks and resettling the loose ones

(6) Reinforcing weakened bricks with a preparation based on silicic acid ester

(7) Filling the losses in the shaped bricks with coloured mineral mortar

(8) Removing loose dirt and deposits from the surface of the plaster

(9) Impregnation with a preparation based on the nanoparticle colloidal solution of calcium hydroxide

(10) Gluing exfoliated and loosened fragments of the plaster with an injection mass based on colloidal solution of calcium hydroxide

(11) Filling plaster losses with a sand-lime mortar

(12) Securing the plaster surface and consolidating the colouring of the plaster surface with a layer of a lime paint

7.4.8 Conservation Treatment

Necessary works were performed on the finial of the blind window. It was crucial to consolidate the structural break running through the finial and to replace the missing shaped bricks framing the blind window. A deep crevice along the left edge of the blind window was filled. PLM-A mortar (Kremer) was applied in injections. The injection mass flowed from the inner side of the wall. Next, the weakened layers of plaster began to be consolidated and reinforced. Consistently, step by step, CaLoSiL® E25 diluted to 12.5% was injected into cracks, crevices and losses.

The procedure was designed to introduce the reinforcing preparation as evenly as possible in order to saturate the whole layer of plaster through to the brick bed. The same spots were saturated repeatedly until the absorption visibly decreased. Some crevices and exfoliations which had been disintegrated more seriously were also saturated with undiluted CaLoSiL® E25. While the cracks and exfoliations were being consolidated, protective mortar wraps were laid on the edges of the plaster losses. The bands consisted of CaLoSiL® this should be hyphen based mortar mixed with a natural aggregate of rinsed and sieved river sand. The granulation of the aggregate was 0.2–0.3 mm. The ratio of the binder to the aggregate was 1 : 3. The mortar was coloured with natural pigments. The

(a)

(b)

Figure 7.65 Introduction of nanolime into the structure of the plaster through cracks, fissures and losses.

mortar wraps protected the plaster from crumbling and reduced any leakage of the preparations introduced into the plaster.

Scales and exfoliations of the plaster were filled with an injection mass based on the colloidal solutions of calcium hydroxide and a specifically designed composition of aggregates. In order to increase the liquidity, the preparation was diluted with CaLoSiL® E12.5 and ethanol in the ratio of 3 : 1. After the exfoliated parts of the plaster

Figure 7.66 The filling of losses and the installation of protective bands around split fragments (detached from the bedding).

were glued and stabilised, CaLoSiL® E12. F5 was applied on the whole blind window twice at 6-day intervals. When CaLoSiL® E12.5 was used for impregnation, there was no haze appearing on the surface. After reinforcement and consolidation effect was obtained, plaster losses were filled. The applied mortar was based on pitted lime (Keim's Romanit Sumpfkalk) mixed with a natural aggregate of rinsed and sieved river sand. The granulation of the aggregate was 0.2–0.3 mm. The ratio of the binder to the aggregate was 1 : 3. The mortar was coloured with natural pigments. Finally, the plastered surface of the blind window was covered with a coloured lime whitewash based on pitted lime (KEIM Romanit Sumpfkalk) and natural pigments (KEIM Romanit-Volltonfarben).

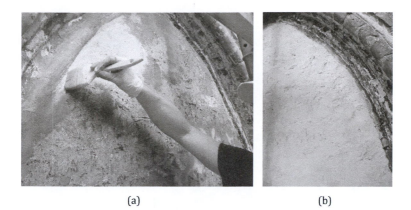

(a)   (b)

Figure 7.67 After the completion of all the necessary technical procedures, the surface of the historic plaster was painted with lime paint.

Preparation consumption was the following:

- Filling fissures, consolidation scales and gluing exfoliations – 1910 mL CaLoSiL® E12.5 [0.14 L/m^2]
- 1st application – 1800 mL CaLoSiL® E12.5 [0.17 L/m^2]
- 2nd application – 2140 mL CaLoSiL® E12.5 [0.21 L/m^2]

7.4.9 Assessment of the Treatment Results

The degraded plaster consolidation method developed on basis of examination results proved to be effective. The preparation penetration obstacle was overcome by diluting the preparation to 12.5 g/L concentration. This also had a positive impact on reducing the occurrence of surface haze. A small addition of hydroxypropyl cellulose practically eliminated the formation of haze. Due to the visible and tested effect of consolidation, only two applications were conducted on the entire surface of the blind window. Cracked and exfoliated areas were treated more times by means of injections with CaLoSiL® E25, CaLoSiL® E12.5 and an injection grout (based on colloidal solutions of calcium hydroxide). Average consumption of CaLoSiL® E12.5 and E25 (for all kinds of treatment) was 0.54 L/m^2. The object is still monitored today. No worrying changes have been observed.

Figure 7.68 Cathedral with the blind window from North-West four years after the completion of treatment.

Figure 7.69   Monastery Rosa Coeli in Dolní Kounice.

7.5 Dolní Kounice*

7.5.1 Introduction

Consolidation of the historical plasters on the second floor of the cloister of the former monastery Rosa Coeli in Dolní Kounice [146] was carried out in 2011. The main aim of the restoration was to salvage the plasters and protect them from further degradation. The restoration above all centred on reinforcing the surface and mass of the existing plasters and reattaching the loose plaster to the foundation layer.

Consolidation of the plaster layers was carried out after an extensive laboratory exploration of the authentic materials gathered from the monastery and from on-site research. This included the testing of selected reinforcement methods that are commonly used in the consolidation of mineral materials and a series of comparisons with commercially available nanosols of calcium hydrate and an assessment of their relative effectiveness. After analysis of the results, the most suitable reinforcing agent for the comprehensive

---

*This section is written by Luboš Machačko and Karol Bayer.

reinforcement of the historical plasters was chosen. Laboratory exploration of plaster materials, tests of reinforcement methods and the restoration process of the west wall of the cloister was undertaken within the framework of the international STONECORE project [147].

7.5.2 Characterisation of Preserved Plasters and Mortars

Three distinct historical plaster layers were detected on the west wall of the second-floor cloister of the monastery. The first layer, preserved in fragments, is a relatively crumbly, rough-grained mortar of pinkish hue, containing tiny particles of carbonated lime. The pinkish hue is due to the presence of iron elements (oxides and hydrated oxides) in the sand that was originally used to produce the plaster. Analysis showed that the binder used in the mortar is most probably white aerated lime. The plaster is relatively rich in binder. The main soluble component (in terms of the content of compounds of carbonate – calcium carbonate formed by carbonation of lime) makes up almost 45% of the substrate in diluted HCl (1 : 4). The aggregate used, was in the form of silica sand containing, alongside quartz, other siliceous particles and fine mineral chips. The distribution of particle size is relatively wide; the majority of grains are between 0.25 mm to 4.0 mm, with none being larger than 30 mm.

This material is probably the original mortar from the end of the 12th century, and is spread onto the masonry in many places.

The second layer has been preserved in only a few places and is a light grey plaster with a smooth surface. The binder is white aerated lime containing the same sand that has been used as aggregate in the masonry mortar. Likewise, this plaster is relatively rich in binder, with the proportion soluble in diluted HCl amounting to almost 43%. On the lower, right section of the wall there is a larger preserved area of the same plaster (about 0.9 m^2) with a particularly smooth surface and relatively marked entanglements. Its character suggests that the materials in this layer possibly date back to modifications made to the monastery in the second half of the 14th century.

Figure 7.70   West wall before treatment.

Figure 7.71   West wall, mapping of preserved plasters: Pink: 12th century, Blue: 14th century, Green: 18th century.

Figure 7.72   West wall after treatment.

The third layer consists of a rough-grained, extremely corroded, greyish layer of plaster. It contains larger particles, between 2 and 3 cm, of un-amalgamated carbonated lime. The proportion of calcium carbonate in diluted HCl is almost 40%. Due to the increased content of un-amalgamated lime particles, the mortar can be said to have a higher soluble proportion compared to that of the lime used as an actual binder in the plaster. The larger un-amalgamated particles tend to act as more of a binding component. The aggregate was silica sand, containing, alongside quartz, other siliceous particles and fine mineral chips. Once again, the distribution of particle size is relatively wide; the majority of the grains are between 0.25 mm and 4.0 mm, as in the previous case, none being larger than 30 mm. This plaster layer covers an approximate area of 40 to 45%

Figure 7.73 West wall, a detail of two layers of preserved plasters: (left) 12ᵗʰ century and (right) 18ᵗʰ century.

of the wall. Most probably, this plaster dates back to modifications carried out on the monastery during the 18ᵗʰ century.

7.5.3 Condition of the Preserved Plasters

Prior to reinforcement, the condition of the existing plasters could have been classified as derelict. The plaster on the west wall of the cloister was only partially preserved. It was badly damaged and eroded with an extremely crumbly surface and in certain places detached from the foundation layer. Dust and grime were evident throughout the area of the surface. There was localised evidence of rainwater ingress. In some places there was a thin and extremely fragile surface layer under which the mass of the plaster had completely disintegrated. Furthermore, in places there was evidence of cracks and fissures.

7.5.4 Tests of Consolidation Agents: 1ˢᵗ Stage

Initially, six commercially available consolidation agents were tested in order to select the most effective for the general reinforcement of the plaster. The candidate agents were applied by spraying onto

Table 7.14   Overview of tested consolidation agents

Product designation/ manufacturer	Type	Test area	Amount of consolidating agent
CaLoSiL® E50 (IBZ-Salzchemie, D)	Nanosols of calcium hydroxide in ethanol	16 × 25 cm	200 mL (2 cycles of 100 mL)
CaLoSiL® IP25 (IBZ-Salzchemie, D)	Nanosols of calcium hydroxide in isopropyl alcohol	15 × 25 cm	200 mL (2 cycles of 100 mL)
KSE Steinfestiger 300 (Remmers, D)	Ethyl ester of silicic acid	16 × 25 cm	150 mL
KSE Steinfestiger 510 (Remmers, D)	Ethyl ester of silicic acid	16 × 22 cm	150 mL
Porosil® RZ (Aqua, CZ)	Ester of silicic acid	17 × 29 cm	300 mL
Porosil® ZTS (Aqua, CZ)	Silicic solution	17 × 33 cm	150 mL

test surfaces until saturation point was reached, that is, to the point when the plaster can absorb no more consolidation agent.

Within the framework of the tests, three basic types of consolidation agent were compared – those based on silicic acid esters (3 commercially produced types), those based on colloidal aqueous solutions of silicic acid (1 commercially produced type) and those based on nanosuspensions of calcium hydroxide in alcohol (2 commercially produced types).

7.5.5 Subjective Assessment of the Consolidation Agent Tests

The subjective assessment of the effectiveness of the consolidation was carried out one month after application of each agent. The tests carried out on the reinforced surfaces were as follows:

- Visual assessment: Evaluating optical changes to surface of consolidated material after application of consolidation agent.
- Scratch test: Subjective evaluation of rate of reinforcing based on comparing crumbling of consolidated plaster with referential plaster prior to consolidation – scratch carried out with scalpel. The depth of penetration of the consolidation agent was also assessed indicatively in this manner, and is directly

proportional to the depth and rate of consolidation of the material.

- Estimation of the penetration depth of each consolidation agent.

The penetration depth achieved by all the agents into the heavily corroded plaster was good. The deepest penetration was detected in the case of the organic silica consolidation agents (CaLoSiL® E50 and CaLoSiL® IP25 penetrated to a depth of 11–14 mm, KSE 300 (Remmers, Germany) and KSE 510 (Remmers) to a depth of 15–18 mm, Porosil®RZ to a depth of 15–17 mm, Porosil®ZTS to a depth of 10–14 mm.). The consolidating effect with a single, respectively double application in the case of nanosols of calcium hydroxide in alcohol, was relatively low. This could be due to the fact that the plaster on specific areas is badly damaged, and it would be necessary to apply the consolidation agent in multiple cycles. The consolidation agent based on colloidal silica was an exception but, in spite of having a greater tendency to consolidate the plaster surface, it had a relatively low consolidating effect below the surface.

Figure 7.74 West wall, tested consolidants: 1 – CaLoSiL® E50, 2 – CaLoSiL® IP25, 3 – KSE Steinfestiger 300, 4 – KSE Steinfestiger 510, 5 – Porosil®RZ, 6 – Porosil®ZTS.

Consolidation agents based on silicic acid esters and on colloidal aqueous solutions of silicic acid have a tendency to accentuate the shade of the plaster. Conversely, consolidation agents based on nanosols of calcium hydroxide form a fine white haze on the surface of the consolidated plaster. The most likely cause of this white haze is the formation of an extremely thin layer of calcium carbonates on the plaster surface.

Objective assessment was carried out using micro-destructive methods – drilling resistance and the so-called 'peeling test' – 17 months after application of the consolidation agent (see Chapter 2).

7.5.6 Tests of Consolidation Agents: 2nd Stage

The next stage of the STONECORE project concentrated on the testing and in situ application of the CaLoSiL® E25 nanosuspension (nanosol of Ca(OH)$_2$ in ethanol). This consolidating agent, from the CaLoSiL® range, produced very good results during laboratory tests and, in some cases, produced superior results to those obtained from tests carried out earlier with CaLoSiL® E25 and CaLoSiL® E50. For this reason, further consolidation tests were carried out on five, 40 cm × 40 cm test areas using spray application. During these tests special attention was paid to the amount of nanosol absorbed by the plaster, the duration of application and the formation of white haze. The influence of pre-wetting the plaster with alcohol on the speed and depth of penetration was also monitored. The consolidation agent was applied to the test area in spray form until the saturation point was reached in one to four cycles.

7.5.7 Subjective Assessment of Tests with CaLoSiL® E25 Nanosol

The depth of penetration into the plaster was detected using a 1% solution of phenolphthalein immediately after application of the nanosol. This depth was measured at four points in each test area. The reinforcing effect was tested subjectively one month after application of the nanosol. Testing was carried out mechanically (tapping, prising, scratching). Consolidation was similar on all test

Table 7.15 Overview of tests of consolidating plaster with CaLoSiL® E25

Test area	Dimensions of area [cm]	Number of cycles	Amount of consolidating agent [mL]	Duration of application [min]	Depth of penetration (max.) [mm]	Pre-wetting with ethanol yes/no
Square no. 1	40 × 40	4	200	5	5–7	No
			110	2	7–10	No
			270	8	18–20	No
			220	15	18–20	No
Square no. 2	40 × 40	1	200	5	7–10	Yes
Square no. 3	40 × 40	3	200	7	7–10	Yes
			110	2	12–15	Yes
			280	15	18–20	Yes
Square no. 4	40 × 40	3	200	3	5	No
			140	2	7–10	No
			280	15	12–15	Yes
Square no. 5	40 × 40	2	280	5		Yes
			280	5		Yes

areas. Test area no. 1, where the highest volume of consolidation agent was applied in the highest number of cycles, displayed the most promising result. This verified the hypothesis that the reinforcing effect generally improves when there is an increase in both the amount of nanosol used and the number of application cycles.

The formation of white haze is very low or imperceptible, with agglomeration of particles of nanosol only on the surface of the plaster in places where dark crusts are present and there is insufficient penetration of the consolidation agent into the mass of the plaster itself. The effect of pre-wetting the dry material on the depth of the penetration was found to be negligible.

7.5.8 Assessment of Consolidation Tests

An assessment of the reinforcing effect of all the potential consolidation agents was carried out 17 months (CaLoSiL® E25 after 2 months) after their application using the 'peeling test', drilling resistance and by measuring absorption capacity. These measurements confirmed an increase in the strength of the surface of the plasters in all cases after application of the consolidation agents. The most promising reinforcing effect was registered with agents based on silicic acids – Porosil®RZ and Porosil®ZTS. The disadvantage of these agents is their extreme reinforcing effect on the surface compared to their relatively low consolidating effect below that surface, and the low or almost negligible absorption capacity of the treated plaster.

A sufficient reinforcing effect was registered in the case of KSE Steinfestiger 300 and KSE Steinfestiger 510. The drawback of these agents is again the very low or negligible absorption capacity of the consolidated material even 17 months after treatment. The result achieved with CaLoSiL® E25 was very good, and this application was actually beneficial in the uniform strengthening of the plaster. A disadvantage of this method of consolidation could be the necessity for several consolidation cycles in order to reach an adequate level of consolidation, resulting in the risk of the formation of white haze on the surface of the material. In this case a relatively low concentration (25 g/L) helped to reduce the formation of white haze to a minimum. Moreover, the particle size (50–250 nm) guaranteed satisfactory penetration into the pore system of the consolidated material. After assessing all the tests, the nanosol of calcium hydroxide in ethanol – CaLoSiL® E25 – was chosen as the most suitable agent for the structural reinforcing of the studied plasters.

7.5.9 Overall Reinforcement of Plasters

After assessing all the tests of the above-mentioned consolidation agents, the nanosol CaLoSiL® E25 was chosen for the final reinforcement of the plaster. The reasons for its selection included: its similarity to the plaster, the adequate reinforcing effect, a suitable particle distribution within the material, and that it has the least effect on the physical properties of the plaster of all the agents tested.

The plaster layers were consolidated by spraying the nanosol on in three cycles. In order to achieve the optimal consolidation possible it was necessary to prevent the remigration of the agent to the surface of the plaster. This phenomenon would result in a lower rate of consolidation of the plaster and the formation of

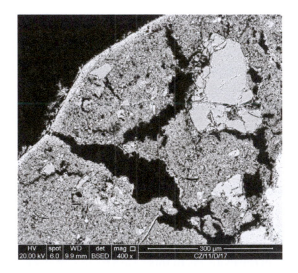

Figure 7.75 SEM-BEI: plaster before treatment, calcium carbonate layer on the surface, pore structure of plaster with no binder. Photo: E. Mascha.

Figure 7.76 SEM-BEI: plaster after treatment, pore structure of plaster with binder (calcium carbonate layer on a surface of a pore). Photo: E. Mascha.

white haze on its surface. Therefore, the consolidation agent was applied selectively through gradual spraying in the morning and late afternoon. The whole process was monitored by a qualified restorer.

7.5.10 Formation of a White Haze

A possible negative effect of applying nanosols for the consolidation of mineral materials is the formation of white haze. This haze is a result of the undesirable agglomeration of calcium particles on the surface of the consolidated material. This agglomeration can be the result two different phenomena. The inadequate penetration of the nanosol into the material, where the pore system with pores of smaller dimensions or inadequately connected, and a surface sealed by a layer with a low absorption rate (crusts, impurities, paints) inhibit the rate of penetration into the material. Alternatively, the formation could be due to the effect of remigration of the calcium particles of the nanosol from deep in the material to its surface. This is a result of rapid or intensive evaporation of solvents from the material.

Table 7.16 Overview of process, condition and results of plaster consolidation with CaLoSiL® E25. (T– Temperature, RH– relative humidity)

Cycle	Date	Climatic conditions	Consumption treated	Area effect	Consolidating	White haze
1	24.05.2011	T: 21.8°C RH: 60.7%	47.5 L	27.9 m²	Increased but inadequate strength	None
2	08.06.2011	T: 22.3°C RH: 73.9%	17 L	27.9 m²	Increased strength, adequate consolidation on over most of surface	None on most of surface in places very faint
3	29.06.2011	T: 26.8°C RH: 42.7%	6.5 L	Where necessary	Adequate consolidation	In places very faint

Figure 7.77   West wall after the treatment from the left.

In some cases, the appearance of white haze can indicate a low rate of penetration of the consolidation agent into the mass of the consolidated material but also insufficient consolidation of the underlying material. The formation of white haze during the consolidation process can be reduced or eliminated to a certain extent by lowering the concentration of the nanosol and applying it gradually and selectively. Immediately after application, it is possible to reduce the formation of white haze by 'stippling' the affected surface with a swab moistened in demineralised water.

The formation of a faint white haze was observed on the surface of the second, light-coloured layer of plaster (in total around 0.6 m^2). As the surface of this layer inhibited the penetration of the nanosol, a layer of haze formed – undesirable agglomeration of calcium particles on the surface – due to the inadequate penetration of the nanosol into some sections of the plaster during the third cycle. The haze was partially suppressed by 'stippling' with demineralised water, and the surface, after drying out, was retouched in places with aquarelle (natural iron oxide pigments and water).

Figure 7.78   West wall after the treatment from the right.

7.5.11 Conclusion

The main aim of the restoration of the plasters on the west wall of the cloister of the former monastery at Dolní Kounice was to salvage them physically and prevent their further degradation. Above all, the restoration dealt with reinforcing the surface and mass of the preserved plasters (increased cohesion) and anchoring the detached plasters to the foundation layer (increased adhesion of plasters). The surface and the mass of the preserved historical plasters were successfully reinforced.

The reinforcing effect was determined by the choice of consolidation agent and method of application, but also by the initial alarming condition of the plaster layers – widespread damage caused by degradation of the binder. The restorers' ambition was not to create from the dissolute plasters a compact, abrasion resistant or hydrophobic surface finish to the walls, nor was it their ambition to 'revive' their former strength (something that no longer exists and of which we know little). Conversely, the ambition was to consolidate the existing fragments of preserved plasters whilst preserving their characteristic material properties in order to postpone their demise as long as possible. The reinforcement of the plaster layers of the former monastery Rosa Coeli in Dolní Kounice is the first case of widespread use of nanomaterials in restoration practice in the Czech Republic.

7.6 Xanten Cathedral*

7.6.1 Introduction

Xanten Cathedral (also known as St. Victor's Cathedral), pictured in Fig. 7.79, is a Roman Catholic church in Xanten, a historic town situated in the lower Rhine area of North Rhine-Westphalia, Germany. It is regarded as the largest cathedral between Cologne and the coast. In 1936 it was declared a minor basilica by Pope Pius XI. Even though called a cathedral, the church has never been the seat of a bishop.

The cathedral owes its name to Victor of Xanten, a member of the Theban Legion in the 4th century who refused to offer sacrifices to the Roman gods, and was reputedly executed in the amphitheatre of Castra Vetera. This Roman camp was not far from the present-day town of Birten. According to legend, Helena of Constantinople recovered the bones of Victor and his legion and erected a chapel in their honour. During a modern excavation the existence of a 4th century cella memoriae was discovered; however, investigations showed that it had not been erected for Victor but for the bodies of two other men placed in the crypt at a later date.

The cornerstone of the cathedral was laid in 1263 by Friedrich and Konrad von Hochstaden. Construction work lasted 281 years, and was finally completed when the Holy Spirit Chapel was dedicated in the year 1544. The nave of the cathedral, consisting of five aisles, was built in the Gothic style. In contrast to many other cathedrals of the period, St. Victor's has no ambulatory. Instead, a pair of chapels is connected to the choir, similar to those seen in the Church of Our Lady in Trier. The monasterial library of the cathedral houses one of the most important religious libraries of the lower Rhine. The cathedral was the seat of auxiliary Bishop

---

*This section is written by Ewa Piaszczynski and Nadine Wilhelm.

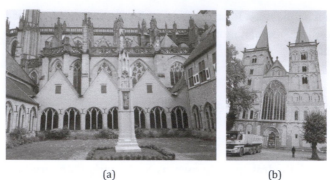

(a)   (b)

(c)

Figure 7.79 (a) South side of the cathedral and the cloister. (b) St. Victor's Cathedral. (c) Arcade of cloister, the two arches on the left side were treated.

Heinrich Janssen, who presides over the lower Rhine region of the Diocese of Muenster. The cloister of the cathedral was built on the southern side of the church. The walls are divided by seven axes on the north and south side and by nine axes on the west and east side. The walls of the cloister are pierced by tracery windows, each divided into three parts by three lancet arches, of which the middle lancet is higher than the outer ones. At the bottom, the tracery is closed by a windowsill. The walls are built of Römer tuff, or brick in the basement area. The columns and tracery consist of Drachenfels trachyte and, in places, of Weibener tuff, Baumberger limestone and brick. The walls were originally plastered both inside and outside.

7.6.2 Damage Characteristics

Large areas of the stone, especially the trachyte and tuff, show extensive signs of damage. The following types of damage occur:

- Scaling and flaking to a depth of up to 1 cm, the scales have only low adhesion to the underlying surface. Under the scale, very small and thin flakes can be found lying parallel to each other, especially on tuff, trachyte and sandstone. Flaking occurs to a depth of 2 cm.
- Changes of colouration to brown and black can be seen on the surface of trachyte, in addition to contamination by microorganisms.
- Every type of stone has signs that original material has been lost as a result of salts, humidity and frost.
- The entire masonry, the ashlars and the joints, are in a bad condition. The cement jointing, which was added during a previous renovation, is partially broken. There are empty spaces beneath the putty. The surface of all the stones has signs of deterioration and sanding.
- Furthermore the surfaces, especially those of sandstone and tuff, are contaminated by black crusts. Cracks which allow water and salts to penetrate can also be found.

Typical damages are shown in Fig. 7.80. The exact location of the various types of damage are given in the mapping, see Fig. 7.81.

7.6.3 Characteristics of the Different Stones

After the mapping, several samples were taken for analysis. Various tests were also made on-site. The following methods were used:

- Drilling resistance measurement
- Peeling test
- Porosity and pore size
- Energy-dispersive X-ray spectroscopy (EDX) and SEM micrographs
- Petrographic investigations

The structure is built of different types of stone.

Figure 7.80 (a) Dark crusting. (b) Loss of material due to flaking. (c) Imperfections. (d) Crack and sanding. (e) Original plaster and paint. (f) Original joint mortar.

7.6.3.1 Römer tuff

This pyroclastic, brownish-grey stone is weakly compacted. In general it consists of strongly altered groundmass and minor amount of minerals (phenocrysts) and lithic fragments. Phenocrysts are represented by feldspar, augite, quartz and opaque minerals. Lithic fragments are mainly basalts and less common pumice.

(a)   (b)

(c)   (d)

Figure 7.81 (a) Types of damage of Axis I. (b) Types of damage of Axis II. Colour code: yellow: unstable surface, sanding; orange: crusting, localised spalling; purple: heavy loss of material due to flaking; red: imperfection and loss of surface due to localised spalling; blue: empty spaces under the surface, with flaking edges and cracks, green: microorganisms. (c) Types of material used Axis I. (d) Types of material used Axis II. Colour code: red: Römer tuff; bright green: Weibener tuff; dark green: Drachenfels trachyte; purple: Baumberger limestone; yellow: brick; lila: plaster.

Table 7.17   Mineralogical composition (vol%) of Römer tuff

Quartz/Feldspar	Augite	Lithic fragments	Opaques	Matrix	Pores
2.5	4.5	22.5	1.5	63.0	6.0

Table 7.18   Characteristics of the crusts, microscopic observation in reflected light

Investigation method	Detected compounds
SEM-EDX	C, Ca, Fe, Si No gypsum
Structural analysis XRD	Quartz (silica) SiO_2
Micro-Raman analysis of the chemical composition Specimen from the stone facing	Quartz (silica) SiO_2 Exposed surface of the specimen: No gypsum Carbon

A detailed analysis of the local stone sample is summarised in Table 7.17. Typical microscopic pictures of thin sections of the stone sample are shown in Fig. 7.82.

The most popular phenomena of weathering are the loss of soft aggregates, scaling, flaking and the rounding off of corners and edges on square blocks. The data given in Table 7.18 gives the characteristics of dense, soot-blackened crusts on Römer tuff.

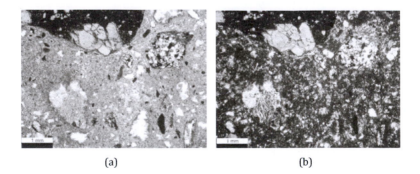

(a)   (b)

Figure 7.82   Römer tuff. sample no. 24, thin section in (a) plane-polarised light (b) cross-polarised light. Photos: W. Bartz.

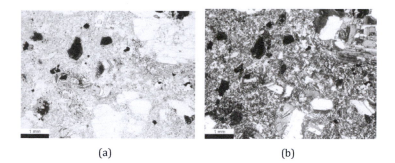

(a)   (b)

Figure 7.83 Drachenfels trachyte, microphotographs of a thin section of sample no. 04 in (a) plane-polarised light; (b) cross-polarised light. Photos: W. Bartz.

Results show that there is no gypsum in the investigated crusts but carbon, calcium, iron and silica. It is supposed that the crust is the result of soot, perhaps of air pollution.

7.6.3.2 Drachenfels trachyte

This compact, whitish-grey stone type is composed of a micro-crystalline matrix (groundmass) with embedded numerous phenocrysts. The matrix consists of microlites of feldspars, pyroxenes and opaques, with sparse sphere, apatite and zircon grains. Generally the grain-size of the matrix constituents is lower than 0.1 mm. The following Table 7.19 gives a summary of the analysis of a stone sample. Figure 7.83 represents a typical polarised-light microscopic picture of Drachenfels trachyte.

Table 7.20 shows the characteristics of a Drachenfels trachyte surface with dense black crust and loose soiling. The most weathering problems are chipping of salidines, scaling and flaking.

Table 7.19 Mineralogical composition (vol%) of Drachenfels trachyte

K Feldspar	Plagioclase	Quartz	Biotite
15.0	1.5	2.5	6.0

Pyroxene	Opaques	Matrix	Pores
2.0	1.5	63.5	8.0

Table 7.20 Characteristics of the crusts, microscopic observation in reflected light

Investigation method	Detected compounds
SEM-EDX	C, Ca, Fe, Si No gypsum
Structural analysis XRD	Quartz (silica) SiO_2
Micro-Raman analysis of the chemical composition Specimen from the stone facing	Quartz (silica) SiO_2 Exposed surface of the specimen: No gypsum Carbon

7.6.3.3 Baumberger limestone

This compact, yellowish-grey stone consists of a biogenic as well as a terrigenous material. The most part of the rock is composed of carbonate phases (sparry calcite and micrite).

7.6.4 Laboratory Testing

Before the consolidants (nanolime dispersions alone or in combination with SAEs) were used, extensive laboratory tests were carried out. The priority was to determine if the consolidants have any negative impacts on the stones.

Table 7.21 summarises the analysis of a stone sample followed by Fig. 7.84 representing a typical polarised-light microscopic picture of Baumberger limestone.

Table 7.21 Mineralogical composition (vol%) of Baumberger limestone

Quartz feldspar	Glauconite	Opaques	Carbonates	Pores
1.5	0.5	3.0	84.0	11.0

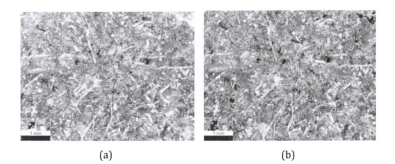

(a)   (b)

Figure 7.84 Baumberger limestone, microphotographs of the thin section of sample no. 5 in (a) plane-polarised light; (b) cross-polarised light. Photos: W. Bartz.

7.6.4.1 Sandwich-form sample

EDX analysis (from sample no. 4) shows a high peak in the Ca and Si distribution. Figure 7.85b,c,d shows that the consolidant probably can be taken as amorphous calcium silicate. CaLoSiL® (alkaline environment) is a hydrolytic catalyst of polycondensation and accelerates the hydrolysis of ethoxy groups.

The texture of the consolidant can be seen in the broken bubble structures. P. J. Koblischek found that the broken bubble structures of KSE gels are the result of incompletely hydrolised monomer structures [148].

Typical gel structures obtained by the consecutive treatment of different stone structures with CaLoSiL® and KSE are given in Figs. 7.85–7.87. In all cases, amorphous silica gels were formed, in which calcium ions are incorporated. Films have been formed bonding the particles together. The SEM micrographs reveal that the consolidants do not form a smooth film but rather cauliflower-shaped structures that demonstrate good adherence to the grains.

7.6.4.2 Cubic samples

It is well known that the distribution of the consolidants inside a deteriorated stone determines the success of a conservation treatment. In order to simulate a situation in which loose stone powder has been bound on healthy stone, cubic stone samples

(a)

(b)

(c)

(d)

Figure 7.85 Römer tuff: sample no. 4, fine powder 0–1 mm with stone. Combination: CaLoSiL® E25 and KSE 100. SEM micrographs (a) magnification factor 1500 (Photo: E. Mascha); (b) magnification factor 4400; (c) magnification factor 9220; (d) magnification factor 25560. Photos: D. Kirchner.

have been prepared containing a hole in the middle and filled with loose stone flour. The powder was consolidated with different combinations of CaLoSiL® products alone or in combination with KSE 300. The consolidants were applied up to three times with a syringe until saturation. Afterwards the samples were stored at RH 75% for three weeks.

For characterisation the whole compounds including natural stone were vacuum-embedded in epoxy resin (Araldite® 2020). Polished sections were produced perpendicular to the surface of treatment. The polished cross sections were coated with carbon

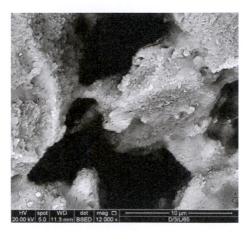

Figure 7.86 Drachenfels trachyte: sample no. 1/65, powder 0–1 mm with stone. Combination: consolidation 2x with CaLoSiL®, 1x with KSE 300; SEM micrographs, magnification factor 12000. Photo: E. Mascha.

Figure 7.87 Baumberger limestone: Sample 5/67, powder 0–1 mm with stone. Combination: consolidation 2x with CaLoSiL®, 1x with KSE 100; SEM micrographs, magnification factor 5000. Photo: E. Mascha.

and studied by SEM (Philips XL 30 ESEM, 20 kV, high vacuum, back-scattered electron [BSE] detector fitted with an energy-dispersive X-ray analyser (Link-ISIS). The SEM micrographs taken at low magnification were fitted together by use of Photoshop® in order

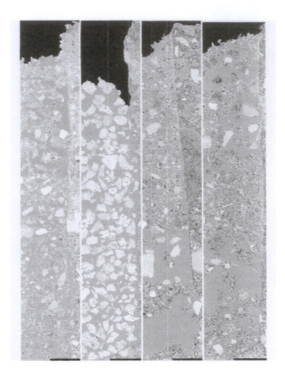

Figure 7.88 Römer tuff: cubic sample no. 77, stone with powder 0–2 mm. SEM-BSE micrograph from left to right: powder 0–2 mm, KSE 300; powder 1–2 mm, KSE 300; CaLoSiL® E25 and E50; CaLoSiL® E25 and KSE 300. Photo: E. Mascha.

to cover the whole sample diameter. Pores were edited in pseudo colour (see Fig. 7.88) in order to ease their visibility and to allow a comparison of the different treatment methods. A homogeneous consolidation effect, and a good fusion of the aggregate filling with healthy stone, were obtained by CaLoSiL® E25 followed by CaLoSiL® E50, respectively CaLoSiL® E50 and KSE 300.

When KSE 300 is applied only, it penetrates through the loose aggregates, but is unable to bind them together. The distances between the aggregates are too big. The combination of the two consolidants remains confined to the part of the stone that has been deteriorated, and does not penetrate into the intact part. It bonds large and small grains together. This method is suitable for stones that have areas of deterioration such as flaking and scaling up to several centimetres, especially tuff, trachyte, Baumberger limestone and calcareous stones.

7.6.5 The Conservation Concept

Based on the results of the pre-investigations, the following conservation concept was developed:

- Cleaning
- Compresses to reduce salts
- Consolidation
- Application of injection grout
- Reconstruction
- Retouching

The surface was cleaned with soft brushes and contaminants such as dust, soot and guano were removed. The microorganisms were removed with a scalpel and a mixture of nanolime and hot water. The cement joints were carefully removed with hammer and chisel without damaging the ashlars. After cleaning, desalination was carried out. All areas that were in contact with cement joints were moistened with water and treated with Arbocel® 300, applied in the form of compresses which were removed after one day. They showed light yellow discolourations, indicating that the salts have been successfully removed.

All stones were consolidated separately. To achieve better penetration, the cracks were moistened with water 24 h prior to the application of CaLoSiL®. Each stone was treated one to eight times with CaLoSiL® E25, depending on the absorption behaviour. The nanolime dispersion was injected with a syringe until saturation. Repeated treatment was realised after evaporation of the alcohol.

In the next step, cracks and empty spaces were filled with an injection grout based on CaLoSiL® paste like [149]. When it had dried, the stones were additionally treated with silicic acid ester. Tuff stone, brick stone and Baumberger limestone were treated with KSE 100; the trachyte with KSE 300. All materials were applied with a syringe. Afterwards the masonry was climatised with wet cloths.

The amounts of consolidants brought into selected stones are given in Table 7.22. Finally, missing parts of the ashlars were

Achse I

Figure 7.89   Axis I, location of tuff elements.

replaced with repair mortar. A special formulation was developed for every different stone variety. The mortar was then retouched so as to blend into the original surface. To achieve this, thin layers of lime paint based on white lime hydrate (produced by Kalk Concept, Siegburg, Germany) were applied without changing the structure of the surface. The location of each element is given in Fig. 7.89.

Alongside the tests, another possible method of application for consolidation was tested in the cloister of Xanten Cathedral: consolidation under vacuum. The consolidant flows through hose lines onto the damaged area, which is covered with foil, sealed and placed under a vacuum. This method was developed by Mr. Matthias Steyer (Dipl.-Rest.) from Niederhausen, Germany (see Fig. 7.90).

The blue areas in Fig. 7.91 (characterising the presence of calcium containing compounds) show that vacuum infiltration has resulted in deep penetration of the stone.

Table 7.22   Amounts of consolidant that was brought into each stone. Location: Axis I, Row 1

Stone	Total amount CaLoSiL® [mL]	Silicic acid ester KSE 100 [mL]
1	213	135.5
2	102	100
3	435	200
4	212	290
5	213	130
6	296	100
7	227	130
8	363.5	140
9	229	90
10	263.5	90

<table>
<tr><td>(a)</td><td>(b)</td></tr>
</table>

Figure 7.90   Vacuum infiltration by the method developed by Mr. M. Steyer: (a) sealed surface and (b) vacuum pump.

Figure 7.91   Drachenfels trachyte: Consolidation with CaLoSiL® E25 under vacuum. Photo: E. Mascha.

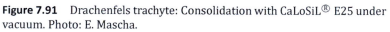

(a)   (b)

Figure 7.92 Restauration: (a) Consolidation with infusion kit. (b) Application of injection grout.

(a)   (b)

Figure 7.93 Restauration: (a) Detail before restoration. (b) Detail after restoration.

7.6.6 Assessment of the Conservation

Comprehensive investigations have been realised to assess the achieved consolidation, selected results will be presented in the following. The homogeneous distribution of the consolidants in Römer Tuff could be proven by SEM investigations. As visible in Fig. 7.94, the consolidants have formed homogeneous layers within the porous stone structure. The injection grout fills the fractures homogeneously. Similar results have been observed in Baumberger limestone. In Fig. 7.95a it is clearly visible that both consolidants build bridges between the grains. The formed layers result in additional strengthening, without filling the pore space completely.

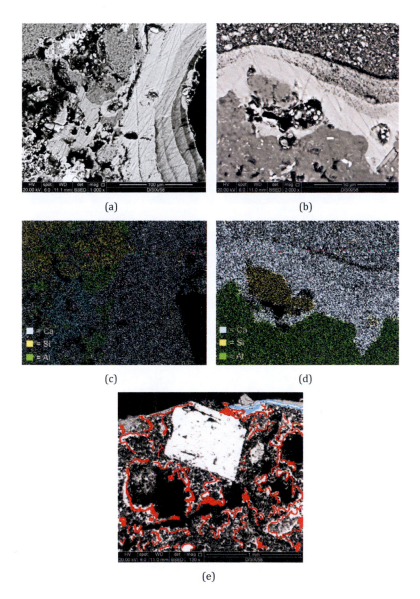

(a)

(b)

(c)

(d)

(e)

Figure 7.94 Römer tuff, SEM-BSE micrographs after conservation – sample no. 58. (a) Sample no. 58, distribution of injection mortar in the stone, magnification factor 1000. (b) Distribution of injection mortar in the stone, magnification factor 2000. (c, d) X-ray distribution of Ca and Si (consecutive application of nanolime and tetraethyl orthosilicate [TEOS]) on the sample. (e) Pore structure, with newly developed layer of consolidants (formed by the reaction of CaLoSiL® with KSE) highlighted in red. Photos: E. Mascha.

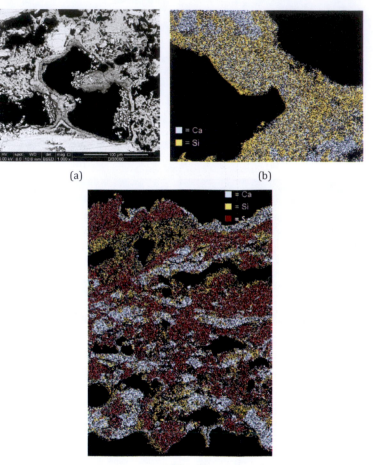

(a) (b)

(c)

Figure 7.95 (a) SEM-BSE micrographs after conservation of Baumberger limestone – sample no. 60. (b) X-ray showing distribution of Ca and Si (consecutive application of nanolime and TEOS), magnification factor 1000 – formation of bridges by means of consolidant in the stone. (c) X-ray showing the homogeneous distribution of consolidants (successive treatment, two times with CaLoSiL® E25 and once with KSE) on the stone. Photos: E. Mascha.

(a)

Figure 7.96 Drilling resistance measurement before and after conservation: sampling points.

The results of the investigations show that both consolidants are able to distribute themselves inside the stone and build bridges between the grains. They are formed by the used consolidant CaLoSiL® E25. The particles of the KSE stick to the bridges.

The achieved mechanical strengthening is demonstrated by the significant increase of the drilling resistance in treated areas, see Fig. 7.96. The data for the left arch are given in Fig. 7.97. Penetration depths up to 15 mm were achieved. This was connected with a significant enhancement of the surface stability, as visible by the results of the peeling tests given in Figs. 7.98–7.100.

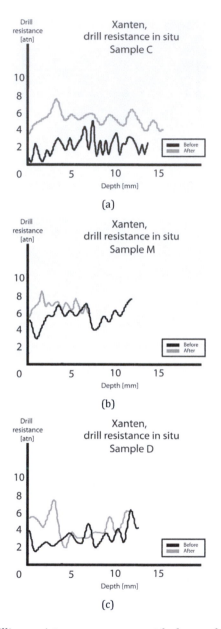

Figure 7.97 Drilling resistance measurement before and after conservation: (a) Römer tuff – sample C; (b) Baumberger limestone – sample M; (c) Drachenfels trachyte – sample D.

(a)

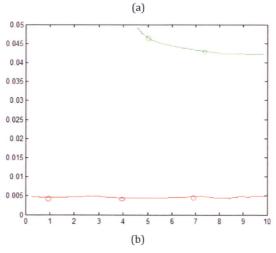

(b)

Figure 7.98 Peeling test, Römer tuff, sample no. XT1. (a) Location of sample no. XT1. (b) Mass loss in [g] before (green) and after (red) consolidation.

(a)

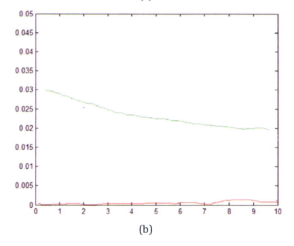

(b)

Figure 7.99 Peeling test, Drachenfels trachyte, sample no. XT2. (a) Location of sample no. XT2. (b) Mass loss in [g] before (green) and after (red) consolidation.

(a)

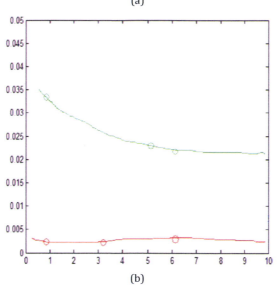

(b)

Figure 7.100 Peeling test, Baumberger limestone, sample no. XB4. (a) Location of sample no. XB4. (b) Mass loss [g] before (green) and after (red) consolidation.

7.6.7 Summary and Conclusions

All laboratory tests and the practical applications have shown that the combination of nanolime dispersions with silicic acid esters, for example, their successive use, allows the consolidation of stones and mortars which are difficult to treat with conventional methods and materials.

Although the nanolime dispersions contain extremely fine particles, they are often unable to penetrate into stones having pore systems with openings <1 μm. In the most materials, however, good penetration is given. When a first structural stabilisation is required, up to six treatments with CaLoSiL® E25 can be realised without resulting in restrictions in the subsequent treatment with silicic acid esters.

The consolidation with silicic acid ester should be carried out between one and eight days after the application of the nanolime dispersions. Due to the higher content on SAE, the mechanical strengths of materials treated with KSE 300 are significantly higher than of those treated with KSE 100. A first post control after five years showed that all treated structures were in perfect condition (see Fig. 7.101). The combination of CaLoSiL® and silicic acid esters is seen as a new, favourable conservation method, especially for the structural consolidation of stones showing spalling and flaking (tuff, sandstone, limestone, brick, trachyte).

(a)   (b)

Figure 7.101   (a) Masonry, axis I, 5 years after the conservation (detail). (b) Masonry, axis II, 5 years after the conservation (detail).

Figure 7.102   St. Martin in Dahlem-Schmidtheim.

7.7 Church Schmidtheim*

7.7.1 Introduction

The following documentation is about a window made of concrete and glass. It is part of the church in Dahlem-Schmidtheim, built in 1960 (Fig. 7.102). The window shows a stylised picture of the Lamb of God which is shaped by open-worked concrete and coloured glass. The window consists of eight concrete elements. Each element contains triangular and square glass shapes. The coloured glass is fixed with mortar. The original surface is preserved and has a light grey to beige colour. All in all, the concrete has an even surface, only some parts are large-pored and show little holes. The elements are joined with mortar that shows a lighter colour and a coarse-grained structure. The window is shown in Fig. 7.103.

All measures were aimed at a complete preservation of the original material and the original shape of the window. Damages typical for concrete corrosion have been found. The damages are

---

*This section is written by Ewa Piaszczynski.

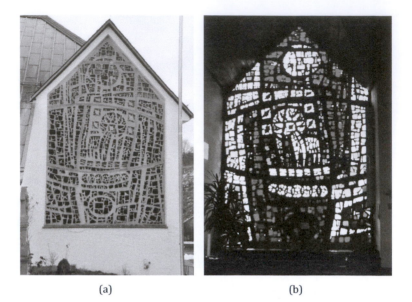

(a)   (b)

Figure 7.103 Window of St. Martin in Dahlem-Schmidtheim. (a) View from outside. (b) View from inside.

indicators for the degree of deterioration to be expected over the next years. A conventional refurbishment would entail uncovering the reinforcement bars, opening the cracks, applying anti-corrosives and then replacing the missing parts. This method, however, would destroy about 70% of the original concrete. Nanolime dispersions are able to prevent further corrosion because they form a highly alkaline environment around the corroded iron. Due to their low viscosity and high flowability they can be applied through cracks and gaps.

7.7.2 Damage Characteristics

At first impression, the window was in a good condition. But on many places, the covering of the reinforcement bars was critically thin, sometimes less than 2 cm (average carbonation depth). All concrete elements show cracks, which can lead to extensive damage.

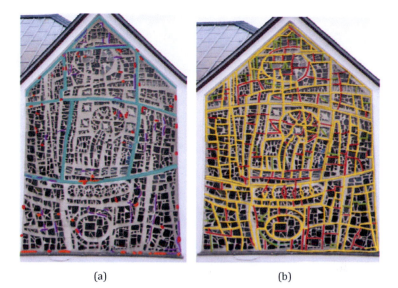

(a)   (b)

Figure 7.104 St. Martin church, Dahlem-Schmidtheim. (a) Mapping of damage. orange: voids; red: uncovered iron bars; purple: cracks; turquoise: joint. (b) Mapping of thickness of concrete. Yellow: less than 2 cm; red: between 4–2 cm; green: more than 4 cm.

The window is contaminated by microorganisms, but the degree is lower than usual on concrete buildings. Furthermore, it shows loss of original material caused by the thin or even missing coating of the reinforcement bars and the resulting splitting due to corrosion. The glass was mostly in a good condition except that some small pieces have broken off. The cracks in the concrete and the holes at the edges of the glass allow the penetration of moisture, which intensifies the corrosion and the splitting due to the freeze-thaw-change and the expanding rust. The joint between window and masonry was filled with silicone and painted with white paint. This joint, however, was very brittle.

In the case of this object it was important to draw a map of the thickness of the concrete (see Fig. 7.104). As long as the reinforcement bars are protected by enough surrounding concrete, corrosion has not been occurred. But, as the pictures Fig. 7.105 shows, there are many places where the thickness of concrete was

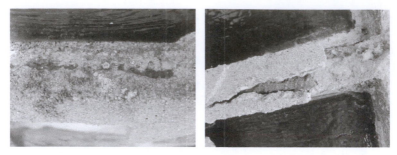

(a)   (b)

(c)

Figure 7.105   Window of St. Martin church Dahlem-Schmidtheim, examples of damage. (a) Microorganisms. (b) Rusty iron bar. (c) Cracks.

too low. A detailed analysis of the stone sample is summarised in Table 7.23.

The deteriorated cement itself occurs as microcrystalline and inhomogeneous mass. It is mostly composed of sub-microscopic crystals of carbonates (micrite), yellowish and weakly transparent, in plane polarised light. Under crossed polars, cementing mass shows high order interference colours, masked by its properties

Table 7.23   Mineralogical composition (vol%)

Quartz	Feldspar	Rock fragments	Others	Cement	Pores
33.0	4.5	29.0	3.0	30.0	0.5

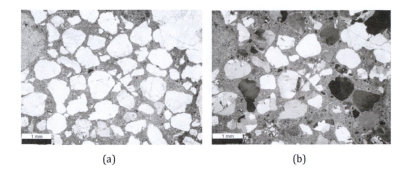

(a)   (b)

Figure 7.106 Microphotographs of the thin section taken in (a) plane-polarised light and (b) cross-polarised light. Photo and investigation: W. Bartz.

visible in plane polarised light (see Fig. 7.106). Numerous relics of Portland cement occur within the micritic mass. They are isometric, of a size not exceeding 0.2 mm. They are composed mostly of small, euhedral crystals of calcium silicates (alite, belite), colourless or yellowish and anhedral brown millerite, located within the interstices. Very small, single fragments of brown millerite grains are scattered within the micritic mass. Locally, the amount of products of cement hydration (C-S-H phases) increases at the expense of micrite, leading to a lowering of interference colours of cement, up to grey, first order.

Especially the horizontal areas are contaminated by microorganisms like *Xanthoria polycarpa*, *Phaeophyscia orbicularis* and *Lecanora caloplaca albescens* (see Fig. 7.107).

These are responsible for granular disintegration and enlargement of concrete porosity on horizontal surfaces. Biological corrosion of concrete is often neglected. It should be noted, however, that a number of physical and chemical processes occurring in concrete are the consequences of the metabolic activity of microorganisms such as bacteria, fungi or algae.

The main cause of damage is the formation of rust on steel reinforcement, which is connected to a significant increase in volume. As a consequence the concrete surface bursts in the area of the rusting steel and the disintegration is accelerated, since water and air get even better access.

(a)   (b)

(c)

Figure 7.107 Microorganisms (a) *Xanthoria polycarpa*, (b) *Phaeophyscia orbicularis* and (c) *Lecanora caloplaca albescens*.

7.7.3 The Conservation Concept

The original material and surface shall be preserved. Therefore, only the damaged areas have to be treated so that the appearance does not change.

The following conservation concept was developed:

- Surface cleaning with water and brushes. Removal of microorganisms with CaLoSiL® E25 and hot water.
- Removal of rusty, uncovered steel bars after consulting a stress analyst. Reinforcement steel, which cannot be removed, will be de-rusted and treated with a rust protective finish.
- Filling of cracks with injection grout based on CaLoSiL® paste like. Prior to this, the cracks will be treated with CaLoSiL® E25 to stabilise the powdering surface inside the cracks and

(a)   (b)

Figure 7.108 (a) Reinforcement bars without treatment. (b) SEM-micrographs of rust, magnification factor 13560. Photo: D. Kirchner.

to enhance the connection between the injection grout and the concrete surface.

- Replacement of missing parts of the concrete with a mineral mortar that matches the surface structure of the original material.
- After the replacement of the missing parts the mortar will be retouched to adjust the mortar to the original colour of the concrete. For this measure, lime colours and colours based on water glass will be used.
- Finish: Finally a very thin translucent protective finish will be applied, if necessary. Its function is to reduce the penetration of water without densifying the surface. The aim is to reduce the water absorption from the outside but to allow condensed water from the inside to evaporate. Otherwise the water will collect in the concrete and cause further damage.

7.7.4 The Conservation Procedure

7.7.4.1 Laboratory testing

In order to characterise the possibility to treat corroded, rusty steel surfaces with nanolime dispersions, complex laboratory investigations have been performed ahead of any applications. Special injection grouts have been developed in order to fill fractures, to stabilise surfaces and especially for realkalisation of damaged zones. The binder of the grouts was always CaLoSiL® paste like, which was mixed with several marble flours (particle size between 0.8 μm and 2.4 μm. The binder aggregate ratio was 1:5 (by mass). Typical properties of a grout are given in the following:

- Water absorption: 25.1 vol%
- Porosity: 37.5 vol%
- Tensile bending strength: $0.04 \, N/mm^2$
- Compressive strength: $1.4 \, N/mm^2$
- E-Modulus: $4.64 \, kN/mm^2$
- Viscosity of the injection grout after one week storage at 50°C: 3600 mPas

To characterise the possibility to fix and stabilise loose reinforcement, a rusty steel bar was put into a hole in the middle of a concrete sample (1 part cement and 4 parts sand; 0–1 mm). Afterwards the space between steel bar and cement was filled with the specially developed injection grouts (see Fig. 7.109).

Figure 7.110 shows a microscopic picture of a sliced concrete sample. In the middle the steel rod is visible embedded in white injection grout.

The samples were put in an outside environment exposed to weather for 2.5 years. To determine also the resistance against salt, the samples were treated with a salt solution 20 times and dried

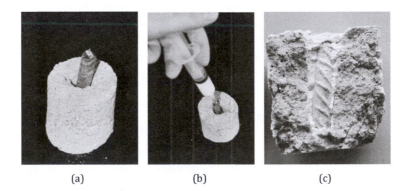

(a)   (b)   (c)

Figure 7.109 (a) Sample of cement with steel bar. (b) Injection. (c) Sample with injection grout.

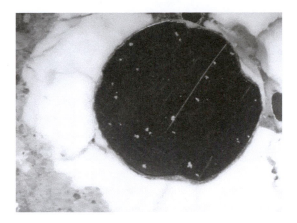

Figure 7.110   Microscopic picture of a sliced concrete sample with steel rod, filled with injection grout. Photo: Zuzana Slížková.

after each cycle. The samples with reinforcement bars embedded in injection grout showed after 20 cycles no signs of damages like cracks, in contrast to the samples without injection grout, which burst open during the test (see Fig. 7.111).

Comprehensive investigations after two years of weathering confirmed the stability of the formed masses as well as the presence of alkaline conditions. The reinforcement is strongly embedded into the matrix formed by the injection grout and no indications of deterioration processes were visible.

7.7.4.2 Conservation treatment

Initially, the surface was cleaned with brushes and water. The microorganisms were removed with CaLoSiL® E25 and hot water (see Fig. 7.112). Tests with other mildew removal agents were ineffective. Broken parts of the window were carefully removed to fix them later. Rusty nails which were embedded in the concrete all over the window were cut out to avoid further corrosion. The cracks were treated with CaLoSiL® E50 to stabilise the surface and to bond loose grains. The consolidation agent was applied with a syringe. In the next step the cracks were filled with the injection grout, based on CaLoSiL® paste like. To achieve a better penetration, the cracks

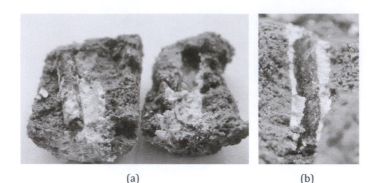

(a)   (b)

(c)   (d)

Figure 7.111 Investigations in weathering. (a) State of the sample after salt test. Where there was no injection mortar, salt crystallisation and crack formation occurred in the sample. (b) Condition of the injection grout bed after salt load test. (c) Steel reinforcement after salt effect, steel reinforcement with injection grout after salt action. (d) Protection layer of injection grout on the steel.

were moistened with water 24 h beforehand the application of the injection grout (see Fig. 7.113).

But it was important that the water could evaporate before treatment with the nanolime dispersion. Drilling resistance measurements before and after the treatment confirm the successive filling with injection grout. Measurements were performed with DURABO measuring instrument in the red marked point illustrated in Fig. 7.114a. This machine drills a little hole in the stone while noticing the needed force in relation to the depth of drilling. In the

(a)   (b)

Figure 7.112 Removal of microorganisms. (a) Cleaning with CaLoSiL® E25 and hot water. (b) Cleaning with scalpel.

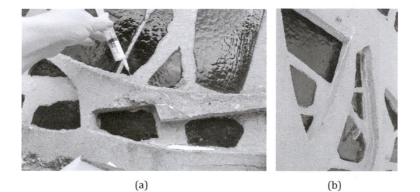

(a)   (b)

Figure 7.113 (a) Filling of cracks. (b) Cracks with injection grout based on CaLoSiL® paste like.

diagram in Fig. 7.114b, the black line shows the drilling resistance before treatment. Clearly visible is a very soft area in a depth of 4–5 mm, which shows almost no resistance at all. This indicates a void between a scale and the underground. The red line in the diagram characterises the drill resistance after treatment with injection grout. The weak area in a depth of 4–5 mm has improved and is obviously filled up by injection grout.

(a)

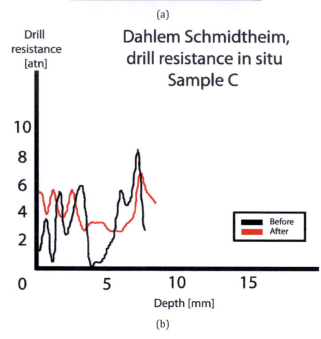

(b)

Figure 7.114 (a) Sampling point marked in red. (b) Drilling resistance measurement before (black line) and after (red line) the treatment.

Uncovered reinforcement steel was cut out where it was possible. The remaining steel bars were freed from concrete and de-rusted. Then they were treated with an anti-corrosive layer, before covering them with new putty. The missing parts were replaced by a mineral mortar whereby the mortar was matched to the original substance, regarding the structure and colour of the surface (see Fig. 7.115). Wherever necessary, rust-proofed stainless steel was fixed in the

(a)   (b)

Figure 7.115 Replenishment of defects: (a) New armouring. (b) Putty (mortar, not yet dry).

local separations to achieve higher stability. The replaced parts were moistened for three weeks. After the replacement of the missing parts the mortar was retouched to match the mortar to the original colour of the concrete. For this measure lime colours and colours based on water glass were used. After that, a translucent protective layer based on water glass was applied. Initially testing areas were chosen to find the right mixing ratio of binding agent and water so that the appearance of the surface would not be changed. Especially the horizontal areas are harmed by humidity. To avoid further damage it was decided to apply a hydrophobic layer onto these areas. The hydrophobic finish Faceo-Oleo HD (Co. PSS Interservice) was used for this measure.

7.7.5 Conclusions

A conservation concept based on the coupled application of nanolime dispersions and special injection grouts containing nanolime as binder was developed and successfully applied. In Fig. 7.116 the restored window is shown before and after the conservation treatment. Strongly alkaline conditions are obtained, guaranteeing a safe realkalisation. Pre-treatment with nanolime dispersions enhanced the adhesion of the injection grout both to the reinforcement and the original concrete structures.

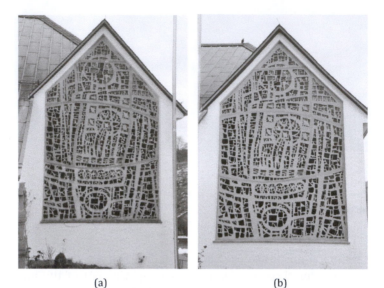

(a) (b)

(c) (d)

Figure 7.116 (a) St. Martin, Dahlem-Schmidtheim, window before restoration. (b) The window after restoration. (c) Detail of the window before restoration. (d) Detail of the window after restoration.

7.8 Aachen Orsbach*

This report deals with the conservation and restoration of the gable of the rectory of St. Petrus in Aachen Orsbach (Germany) (see Fig. 7.117). The gable was constructed in 1764 and consists of marl, a typical construction material in this region. The original mortar consists of marl and lime. During the Second World War the building

---

*This section is written by Ewa Piaszczynski.

Figure 7.117   Rectory before conservation of the gable.

was destroyed. Reconstruction has been carried out after 1945. A later jointing was carried out with cement mortar in the 1960s. The original stones can be clearly distinguished from the replaced ashlars due to the deterioration of their surfaces.

7.8.1 Damage Assessment

On the top of the gable the material is in reasonably good condition, as some ashlars were replaced during a previous renovation. The original stones, however, show extensive damage across their surfaces, which has been mainly caused by weathering. The joints are also in a very poor state. Observable damage includes black encrustation, sanding, flaking, spalling, voids, cracks, discolouration and microbiological contamination. The sharp-edged spalling areas have rough surfaces and some disintegrated zones reach a depth of 5 cm. This deterioration was caused by frost and water, microorganisms, as well as the presence of salts and inappropriate renovations. As result, an extensive loss of original material has occurred and the

(a)   (b)

Figure 7.118 (a) Typical damages of weathered stone. (b) Mapping; yellow: unstable surface, sanding; dark yellow: heavy loss of material due to flaking crusting, localised spalling; orange: heavy loss of material due to scaling; bright red: imperfection; lilac: heavy loss of material due to flaking; dark red: heavy loss of material due to crumbling; dark lilac: voids under the surface; dark blue: voids under the surface with flaking edges; green: microorganism; bright blue: cracks.

Figure 7.119 Mapping of original and replaced ashlars: green: original stone; yellow: substitutes.

stone surfaces are significantly damaged. Prior to the start of the conservation process a mapping profile of the damage was prepared (Fig. 7.118).

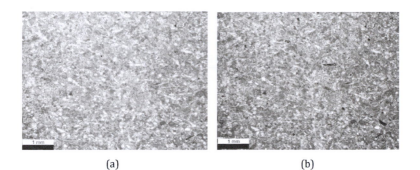

(a)   (b)

Figure 7.120   Microphotographs of thin sections of sample no. 1, in (a) plane-polarised light and (b) cross-polarised light. Investigation: W. Bartz.

7.8.2 Characteristics of Mortar and Stone

The stone is characterised as marl, consisting mostly of carbonates, with minor amounts of terrigenous material. Carbonates occur as bioclasts as well as micritic mud. The latter serves as matrix filling pores between bioclasts, which are represented by ubiquitous spines of sea urchins, ostracodas and less common sponge spicules. A detailed analysis of the stone sample is summarised in Table 7.24. Sea urchins and ostracodas are composed of calcite, whereas sponge spicules are siliceous and composed of opal. In plane polarised light, sponge spicules are colourless to yellowish, whereas in crossed polarised light they are isotropic, dark, contrasting with other rock constituents which are anisotropic. Spines of sea urchins are composed of single crystals of calcite (sparite). In plane-polarised light, they appear as almost colourless crystals, showing fourth order interference colours under crossed polars. Ostracodas are composed of microcrystalline calcite. Typically, there occur patches of microspar inside their moulds. All bioclasts reach up to 0.4 mm in diameter.

Table 7.24   Mineralogical composition of the stone (vol%)

Quartz	Glauconite	Biogenic silica	Carbonates	Pores
2.5	0.5	1.0	94.5	1.5

Table 7.25 Composition of the mortar

Binding agent	CaCO$_3$ %	Acid-resistant contents %	Aggregate
Calcareous hydraulic	57.0	43.0	The mortar contains: 1. Fine grained argillaceous aggregate (grain size <0.1 mm). 2. Great quantity of lime, calcium carbonate was used as aggregate. There are no sand or other quartzite ingredients.

Abundant opaque minerals occur either as larger subhedral to anhedral grains, up to 0.1 mm in size, or very fine-grained homogeneously disseminated framboids, composed of sub-microscopic euhedral to subhedral crystals, presumably pyrite. Few grains of angular to sub-rounded quartz, with size up to 0.2 mm are distributed within the matrix, between bioclasts. In plane-polarised light, quartz grains are colourless, with no cleavage. Under crossed-polars, it shows grey first-order interference colours.

The matrix is composed of weakly transparent micritic calcite, brownish in plane-polarised light. Under crossed-polars, the micritic mass shows high order interference colours.

The mortar is calcareous and argillaceous. Quartz sand was not used. An astonishing number of microorganism could be identified. They covered large parts of the ashlars (see Fig. 7.121). Nitrates and gypsum were present as water soluble salts.

7.8.3 The Conservation Concept

Cleaning: At first the surface was cleaned with soft brushes to remove soiling. The microorganisms have been removed with CaLoSiL® E25 and boiling water. Also, desalination with compresses was carried out.

Removal of the cement joints: Cement mortars were carefully removed mechanically. The original stones should not be damaged during this measure.

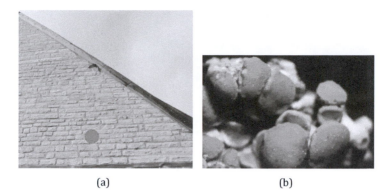

(a)   (b)

Figure 7.121   (a) Location of sample. (b) *Caloplaca saxicola.*

Consolidation: Structurally reinforced and flaking surfaces were consolidated with CaLoSiL® E25 and KSE 300. Injection grout based on CaLoSiL® paste like was used for the filling of cracks.

Reconstruction: The missing parts of the ashlars were reconstructed with mortar and retouched.

7.8.4 Conservation

Due to the poor condition of the gable, a pre-consolidation with CaLoSiL® E25 was required before cleaning of the surface with soft brushes could be realised. The microorganisms were removed with a mixture of CaLoSiL® E25 and hot water. The cement joints were carefully removed using hammer and chisel to ensure the ashlars were not subjected to damage. After these measures desalination was carried out. All areas that were in contact with cement joints were moistened with water and treated with compresses made from Arbocel® 300 and 200. The spent compresses were examined with Multi-lab 540, WTW. Conductivity measurements, showed that before desalination the salt loading was 750 µS, while after desalination the loading had been reduced to 98 µS.

Twenty-four hours before consolidation, the deteriorated stones were moistened with water to facilitate the penetration of the consolidation agent. This was followed by the application of CaLoSiL® E25 using a syringe and infusion kit until the stone became saturated with the consolidation agent. After the stone had

(a)   (b)

Figure 7.122 (a) Consolidation with syringe (b) Consolidation with infusion kit.

dried, the application of CaLoSiL® E25 was repeated six more times. After these multiple treatments with CaLoSiL® E25, a white haze formed on the surface of the stones. This surface haze was cleaned and removed with a mixture of water and ethanol (1:1 vol%). Due to the particularly poor condition of the gable, a further consolidation treatment was necessary. This involved the application of a silicic acid ester KSE 300 and was deemed to be essential as the facade is exposed to weathering (see Fig. 7.122). The additional consolidation with KSE 300 was carried out three weeks after the consolidation with CaLoSiL® E25.

Cracks and voids behind shells were treated with CaLoSiL® paste like, which was applied using a syringe. Surface defects and missing parts of the stones were repaired using lime mortar (Kalk Concept), quartz sand (Strassfeld, yellow fine) and lime pigments. A mixture of coarse lime mortar (Kalk Concept), Rheinesand (0–2 mm) and gravel (0–3 mm) was used in the replacement of damaged joints. To preserve the ashlars, a thin protective layer based on lime pigments (white lime hydrate, Kalk Concept), was applied. The final step involved retouching with lime colours and silicate. In total 55 L CaLoSiL® E25 and 23 L KSE 300 have been used.

7.8.5 Assessment

Four months after consolidation an evaluation of the restoration was carried out.

(a)

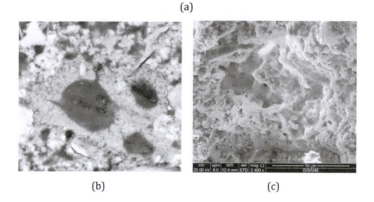

(b)   (c)

Figure 7.123 SEM micrographs of sample no. 48: (a) formation of a secondary pore system; magnification factor 150; (b) bridges created by consolidation agent; (c) after consolidation (twice with CaLoSiL® E25, once with KSE 300), magnification factor 2400. Photos: E. Mascha.

Microscopic investigations showed that both the consolidation agents and the injection grout had distributed widely within the stone and that there was a good adherence to the stone surface (Fig. 7.123a,c). The consolidation agent was found to have formed bridges between different loose areas which can be seen in Fig. 7.123b.

Figure 7.124 shows sample no. F2-21/57 after consolidation with two cycles of CaLoSiL® E25 and once with KSE 300. The micrographs provide evidence for the existence of nanolime and silicic acid ester inside the stone. The formed amorphous silica gel fills the open space in the stone structure and strengthens it.

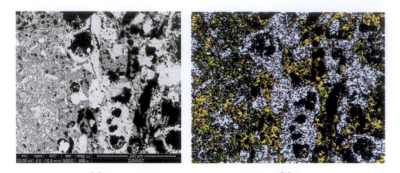

(a)   (b)

(c)

Figure 7.124 (a) SEM micrograph of sample no. F-2-21/57. (b, c) X-ray distribution. Photos: E. Mascha.

CaLoSiL® E25 creates a skeleton on which KSE 300 is deposited. This strengthened the shells and delamination so that the surface of the disintegrated stone will remain stable for a long time. The combination of nanolime and KSE 300 fills large spaces, which KSE 300 would not be able to connect alone. CaLoSiL® E25 creates bridges on which KSE 300 settles, creating a stable reinforcement of the disintegrated parts of the stones. Examination of the microstructure showed that the consolidation agent is deposited on the walls of the large pores and partly fills the small pores and the contact areas between the limestone particles. This means that the consolidation does not change the overall high porosity of the limestone or the relatively coarse porous calcareous material significantly and it may be expected that it will not

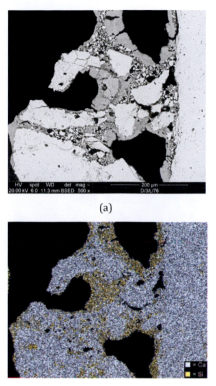

(a)

(b)

Figure 7.125 (a) SEM micrographs of marl samples consolidated twice with CaLoSiL® E25 and once with KSE 300. (b) X-ray distribution of Ca and Si. Photos: E. Mascha.

significantly affect water transport such as water absorption or vapour permeability.

This behaviour was also determined in laboratory investigations with a cubic marl sample (no. 73). Figure 7.125 shows SEM pictures from consolidated marl sample.

The achieved good strengthening was confirmed by drilling resistance measurements (see Fig. 7.126). Especially to a depth of 6 cm the consolidation was absolutely successful and necessary.

Also the peeling test confirmed the excellent consolidation (see Fig. 7.127). The loss of material ahead of the consolidation was 0.4793 g after the consolidation only 0.0049 g.

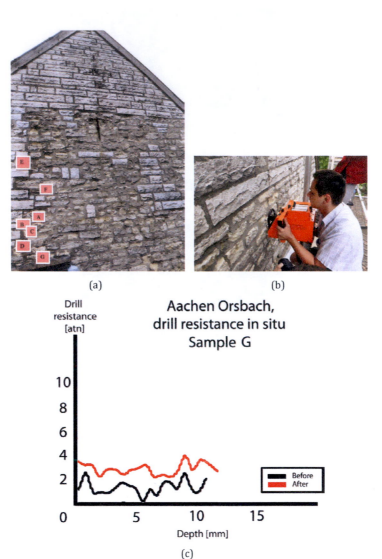

Figure 7.126 Drilling resistance. (a) Sampling points. (b) Drilling resistance measurement – sample G. (c) Result of drilling resistance. Black: drilling resistance – measurement before restoration. Red: drilling resistance after restoration.

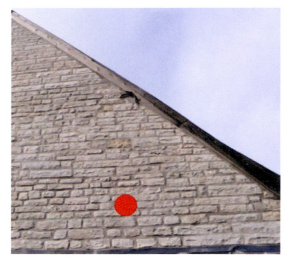

(a)

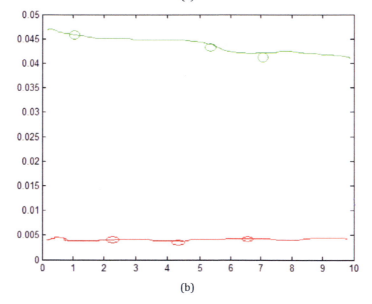

(b)

Figure 7.127   Peeling test. (a) Location of the sample A. (b) Mass loss in [g]: green: before consolidation and red: after consolidation.

(a)   (b)

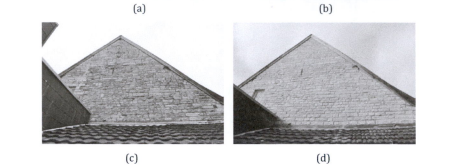

(c)   (d)

Figure 7.128 (a) Detail of the gable before conservation. (b) Detail of the gable after conservation. (c) Gable before restoration. (d) Gable after restoration.

7.8.6 Conclusions and Summary

The successive application of nanolime dispersions and silicic acid esters allowed the successful conservation of materials (marl and related mortars) which are often difficult to treat and to strengthen. Main conclusions are:

- The chosen preparation of nanolime dispersion (CaLoSiL® E25) has shown excellent penetration into marl and mortars.
- The average consumption was $1.20 \, \text{L/m}^2$.
- Consolidation with KSE 300 after fixing with CaLoSiL® E25 caused a significant improvement of the mechanical properties of soft zones.
- The investigations proved that flaking and scaling stones can be consolidated by the successive treatment with nanolime dispersions and silicic acid esters.

- SEM investigations have proven that the nanolime dispersions form adhesive layers on treated surfaces which anchor silicic acid esters. Thus, the application of SAEs becomes possible on surfaces which are normally incompatible to silica gels.
- Drilling resistance measurements have shown a significant increase in strength of the stone structure after fixing with this combination (CaLoSiL® E25 + silicic acid ester). The consolidation agents form stable bridges in the deep areas of the stones.

It is essential that prior to using CaLoSiL® E25 or any other conservation media, pre-investigations should be carried out to identify the most suitable consolidation agent, the most appropriate application method and to determine the required amounts.

7.9 Megalopolis*

Megalópoli is a town in the western part of the peripheral unit of Arcadia, southern Greece. It is located on the same site as ancient Megalopolis. 'Megalopolis' is a Greek word for 'great city'. The city was founded in 371 BC by the Theban general Epaminondas in an attempt to form a political counterweight to Sparta. It was one of the 40 places that were megále pólis (great cities). Megalopolis became the seat of the Arcadian League in 370 BC, which in the 3rd century BC became the Achaean League. It used to be one of the about 20,000 places that had an ancient theatre. In 331 BC, Megalopolis was invaded by the Spartans and there was a battle with the Macedonians who came to Megalopolis' help. The Macedonians defeated the Spartans. In 223 BC, the Spartan king Cleomenes burned down the city, but it was reconstructed by Philopoemen, a Greek General of the Achaean League. The city fell during the medieval times and was reconstructed after the Greek Independence.

In ancient times, the town grew very large. Its theatre, with a capacity for 20,000 people was one of the largest known. In Byzantine and Ottoman times the city was known as Sinanou until

---

*This section is written by Ewa Piaszczynski and Verena Wolf.

Figure 7.129   Ancient theatre of Megalopolis (Greece).

the 19th century. In the mid-1960s, the Public Power Corporation of Greece (PPC S.A.) and the government started the construction of a power plant, which is situated approximately 8 km NW of down town. It was opened in 1969 and provides electric power to southern Greece. A lignite mining area is situated around the plant.

Megalopolis is famous for its ancient ruins, situated to the north-west, including the theatre (see Fig. 7.129). Other landmarks are the Thersileon with 67 pillars and a temple (11.5 m × 5 m, 37 feet × 11 feet). Herodotus reported the ancient belief that the Megalopolis area was a battleground of the Titanomachy. The basis for this was apparently the presence of lignite deposits, which are prone to catch fire in summer and can smoulder and scorch the earth for weeks (Zeus is supposed to have slain the Titans with lightning bolts; see also below), coupled with the presence of fossil bones of prehistoric elephants and rhinoceros. Herodotus informs his readers that the bones of the 'Titans' were exhibited in various places in the surrounding area, at least since the 5th century BC.

7.9.1 Condition Assessment: Damage Phenomena

The theatre was built with dense carbonatic stones of a beige and grey colour and cloudy limonite and iron-compounds (see Fig. 7.130a,b). Residues of plants, algae and mildew can be found (see Fig. 7.130c). The binding media is calcareous, micriteous and homogeneous. In general, the stone has a good resistance to weathering processes and is only lightly bleached. The surface is rough and disintegrated by microorganisms. The stones stand on ground soil and are constantly exposed to weather. Especially in spring, rain enhances the growth of microorganisms and the decomposition of the groundmass of the stone. Formation of cracks and scales are the consequences.

(a)   (b)

(c)

Figure 7.130 Ancient theatre of Megalopolis: seats. (a) Limestone seats. (b) Cracks, crumbling and local separation. (c) Microorganism.

7.9.2 The Conservation Concept

The following conservation concept was developed:

- Cleaning and removal of loose particles
- Removal of microorganisms
- Partial consolidation with CaLoSiL® E25
- Filling of cracks and voids with injection grout based on CaLoSiL® paste like
- Replacement of missing parts and filling of holes with the CaLoXiL® repair mortar
- Retouching with lime colours

7.9.3 The Conservation Procedure

Initially the cracks were cleaned and freed from soiling. The cracks were supposed to be filled with CaLoSiL® injection grout based on nanolime, white lime hydrate and selected marble powders. In a first step CaLoSiL® E25 was applied into the cracks to enhance the connection between the surface of the stone and the injection grout. Afterwards the cracks were sealed with paper and filled with injection grout. After the injection grout had dried, wide cracks and holes were filled with a mortar, which is also based on CaLoSiL®. Finally, the mortar was retouched with lime colours. A total volume of 750 mL injection grout, 300 mL CaLoSiL® E25 and 1.5 kg repair mortar CaLoXiL® was needed.

The injection grout and the repair mortar could fill the cracks and voids and bond the broken pieces together. The materials were easy to handle and could be retouched without problems. Another advantage of the materials is their reversibility.

7.9.3.1 Removal of biological growth

Large parts of the seats were covered with a dense layer of lichens, which have substantially contributed to deterioration processes. In a first step, tests were carried out to recover, isolate and identify the microorganisms. The effect of different biocides was tested on

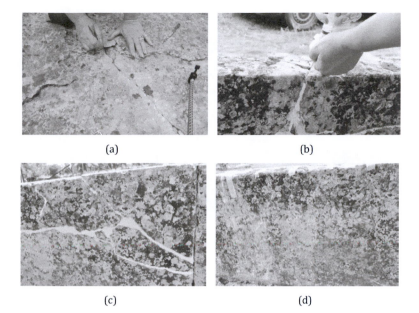

(a)

(b)

(c)

(d)

Figure 7.131 (a) Cleaning of the cracks. (b) Filling of cracks. (c) Filled up cracks with repair mortar. (d) Element after restoration.

selected areas. For that, the test area was sampled using a sterile swab moistened with sterile distilled water. To help differentiate isolates that were simply surviving on the surfaces from those that were colonising the surface an adhesive tape imprint was taken from an area adjacent to where the swab sample was taken. Comprehensive microbiological characterisations carried out at IMSL (UK, Mr. Pete Askew) showed that the microbiology of the exposed area is dominated by lichens (Fig. 7.132).

Despite this, eight distinct fungal strains including again *Penicillium brevicompactum*, were isolated although several other strains producing sterile mycelium were obtained (it is considered likely that they may prove to be the mycobiont component of the lichens present). *Protocooccus sp.* was present along with a small green unicellular algae probably of the genus *Chlorella*. In addition, two unicellular blue green algae were isolated (probably *Cyanothece sp.*

Figure 7.132 Dense growth of lichens around the crack.

and *Gloeocapsa sp.*) as well as a filamentous blue green species (probably *Phormidium sp*). The following treatment procedures were tested in order to remove the lichens:

- Cleaning with H_2O_2
- Physically cleaning
- Washing with *iso*-propanol
- Cleaning with H_2O_2 and two treatments with nanolime dispersions containing 25 g/L $Ca(OH)_2$
- Physically cleaning and two treatments with nanolime dispersions containing 25 g/L $Ca(OH)_2$
- Treatment with nanolime dispersions containing 25 g/L $Ca(OH)_2$, no cleaning

Twenty-four hours after the application of the different biocides, the test areas (Fig. 7.133) were sampled.

The effect of the treatment was determined by standard growth tests in the laboratory. The results are given in Table 7.26. Physical cleaning as well as treatment with H_2O_2 or *iso*-propanol was not sufficient to remove all of the biological growth and the use of nanolime only proved more efficient. However, complete prevention of new fungal growth could not be achieved by a single treatment. As shown in Fig. 7.132 lichens form very dense layers on the

Table 7.26 Results of the treatment with different biocides determined by Mr. Pete Askew, Industrial Microbiological Services Ltd. UK

Sample no.	Treatment	Fungal growth at 9 days	Fungal growth at 16 days	Fungal growth at 1 month
1	Cleaning with H_2O_2	White mucoid growth (some possible *Penicillium*)	White mucoid growth	Mucoid growth
2	Physically cleaning	Grey fungi (some possible *Penicillium*)	Grey fungi (some possible *Penicillium*)	Brown/grey fungus (some possible *Penicillium*)
3	Washing with *iso*-propanol	Off grey fungi (some possible *Penicillium*)	Off grey fungi (some possible *Penicillium*)	Brown fungus
4	Cleaning with H_2O_2 and two treatments with CaLoSiL® E25	No growth	No growth	No growth
5	Physically cleaning and two treatments with CaLoSiL® E25	No growth	Possible *Penicillium*	Possible *Penicillium*
6	Treatment with CaLoSiL® E25, no cleaning	No growth	Green/brown growth	Green/brown growth

Figure 7.133 After application of CaLoSiL®.

surface of the stone. It is considered probable that the nanolime dispersions were not capable of penetrating the lichens completely. The treatment was performed in July at temperatures of around 30°C resulting in a relative fast evaporation of the ethanol. It is assumed that the retention time of the ethanol on the surface of the stone was not sufficient to penetrate and kill the biological growth. This may also be the explanation why the treatment with *iso*-propanol was ineffective. The combined application of H_2O_2 and nanolime resulted in complete disinfection and fresh biological growth could not be detected one month after treatment suggesting that the modified surface is proving more resistant to colonisation that the surrounding weathered stone. In additional tests it could be proven that the dense lichen layer can be removed by combined washing with nanolime dispersions and hot water. This can be realised by using sponges or by spraying. It is important that sufficient time is between the steps to allow the liquids to penetrate deep into the dense lichens layer. Finally, the cleaned surface is treated with a nanolime dispersion containing 15 g/L Ca(OH)$_2$. The resulting calcium carbonate forms a protecting, sacrificial layer.

7.9.4 Conclusions

The combination of nanolime dispersions with injection grouts and repair mortars containing nanolime as binder offers many possibilities to fill cracks and fractures in marble and limestone structures. The conservation materials are fully compatible to the original stone. Their handling is simple. Nanolime dispersions allow also the safe removal of biological growth. This is achieved by the dehydrating action of ethanol in combination with the creation of alkaline conditions by the lime particles.

7.10 Herculaneum*

7.10.1 Basic Description of the Object

The archaeological site of Herculaneum was an ancient Roman town destroyed by a volcanic eruption in 79 AD. Its ruins are located in the vicinity of the Italian city of Naples. In 2009–2010 the trial and study initiative 'Remedial Treatment of Cohesion and Adhesion Failures in Paint Layers at the Archaeological Site of Herculaneum (SGS 09/10)' was organised by the Herculaneum Conservation Project (HCP), a multidisciplinary collaborative partnership between the Soprintendenza Speciale per i Beni Archeologici di Napoli e Pompei ('SANP'), the British School at Rome ('BSR') and the Packard Humanities Institute ('PHI'). Using the wall paintings of the Salone Nero as an example, potential consolidation treatments with inorganic consolidants were tested. Among other institutions the team of University of Fine Arts Dresden (UFAD) has been involved in the SGS-09/10 trials.

The fragmentarily preserved wall paintings of Salone Nero consist of a plane background and raised lines and decorations (see Fig. 7.134). Underneath the paint there are support layers of lime plaster on tuffstone masonry. After previous excavations, parts of

---

*This section is written by Arnulf Dähne, Christoph Herm, and Thomas Köberle.

Figure 7.134   Herculaneum, Salone Nero, southern wall.

the painted plaster were replaced, and fissures and defects were grouted probably using lime cement mortar. All painted surfaces were multiply treated with wax or paraffin solutions or melts. Later, all surfaces were treated with acrylic resin solution. The surfaces of the paintings are variously crossed by fissures. Some areas show discolouration, others have lost colour depth. The damaged areas are localised, and marked by salt efflorescence. Salt analyses have shown the presence of sodium sulphate, gypsum and carbonate. The main deterioration pattern of the paint layer is the extensive delamination of the surface layer from the background paint, locally in the presence of salt efflorescence (cohesion failure: damage type 1). This damage occurs in different scales: There are superficial flakes with thickness below 1 mm and scaling of the upper paint layer with a thickness of 1 to 3 mm. Another deterioration pattern is local splintering of the upper paint layer from the background and separation of raised paint layer (adhesion failure: damage type 2). Generally, causes for the deterioration can only be guessed because extensive research was not possible in the frame of this project. Probably the damages to the paint layers were caused by the combination of the material properties and external conditions. The former porous system of painted plaster has been completely changed by the previous impregnation with wax or paraffin, followed by later treatment with acrylic resin solution. The deterioration process seems to occur between the acrylic insulated surface and the wax-impregnated deeper zones. The current cohesion failure of the background paint (damage type 1) seems to be a result of salt crystallisation, promoted by water infiltration of salts. The salt is crystallising on the surface but also inside the paint layer, probably between the wax-impregnated deeper zone and the acrylic-sealed surface. The lost adhesion and delamination of raised paint layer (damage type 2) may be caused by tension or shear stress between the paint layers. In the background area around the delaminated paint, damage similar to type 1 is present.

The most important conservation action would be the protection against the ingress of water, which can only be achieved through construction measures. Nevertheless, trial tests of the UFAD team concentrated on the following goals:

- Options for the extraction of salts as far as possible.
- Flattening and fixation of loose paint layers.
- Reduction of the former impregnation material where it causes damages. The goal is to reopen the upper part of the paint layer

Generally, the salt efflorescence should be removable by using desalting compresses as usual. With regard to the extraction of impregnation material, investigations with solvent compresses in small test areas should be carried out. Concerning the paint layer damage, the feasibility of swelling and removing the acrylic resin was assessed. Preliminary tests showed that removal of the acrylic resin from the surface is quite possible. The damage potential of the wax or paraffin impregnation and the necessity of its removal – also after a successful refixation of the deteriorated upper paint layer – has still to be investigated. Because the fragile condition of the paint layer is a considerable problem when using any solvent compress or desalting compress, fixation treatment had to be done.

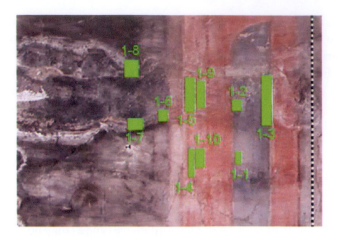

Figure 7.135   Salone Nero, eastern wall, test area TA-1, location of trial tests.

7.10.2 Test Fields

Two sample areas were selected for the test treatments on both damage types. Test treatments on paint layer damages type 1 were carried out in small areas in all the three backgrounds: the black, the red and the blue sections of the South wall (see Fig. 7.135 and Table 7.27). Test treatments on paint layer damages type 2 were carried out on small areas in the base zone (see Fig. 7.135).

7.10.3 Trial Conservation

The details of the conservation tests concerning damage type (1) are given in Table 7.28. An initial test (area 1-1) concerned the possibility of flattening the deformed flakes and scales by pure ethanol. A subsequent test should answer the question, whether flattening also occurs with CaLoSiL®, which also contains ethanol. It appeared that with both liquids, all the thin flakes were levelled spontaneously by capillary effect. Even very heavily deformed or rolled flakes were planed perfectly (see Fig. 7.136).

Thicker flakes or scaling needed to be flattened by finger pressure via a silicone paper. All the paint layer delamination thus could be treated successfully (see Fig. 7.137). The solvent of

Table 7.27 Trial areas, condition (test area 3TA-1)

Area no.	Background section	Damage characteristics
1-1	Blue	Fine flaking paint layer (splitting of the blue paint layer)
1-2	Blue	Larger;scaled delamination of upper part of the blue paint layer, deformed scales; Salt efflorescence
1-3	Blue	Considerable delamination, rolling edges and blistering; presence of efflorescence
1-4	Red	Very fine flaking of a very thin surface layer; rolling flakes
1-5	Red	Flaking of a very thin surface layer in fine and larger scale (<1 mm–3 mm)
1-6	Black	Single large thin scale
1-7	Black	Large thick scale, strong efflorescence
1-8	Black	Very fine flaking of a very thin surface layer, rolling flakes, strong efflorescence
1-9	Red	Flaking, of a very thin surface layer in fine and larger scale (<1 mm–3 mm)
1-10	Red	Very fine flaking of the very thin surface layer, rolling flakes

Figure 7.136 Raking light photo (detail of trial area 1-4): left side: treated with CaLoSiL® – planed flakes; right side: without treatment.

Figure 7.137 Raking light photos of trial area 1-3 before treatment (left) and after treatment (right).

the strengthener was softening the brittle flakes which could be flattened later.

Analysis of the results of these treatments (cf. Table 7.28) showed that the CaLoSiL® nanosol did not effect heavy white hazes on the surfaces. This suggests that the strengthener flowed under the loose surface layers.

Table 7.28 Treatment tests and results (test area 3TA-1)

Area no.	Treatments	Results
1-1	Flattening with ethanol, finger above silicone paper	Easy flattening because of capillarity
1-2	3x CaLoSiL® E25	Deformation without change, no haze formation
1-3	3x CaLoSiL® E25, again 1x after 2h and flattening with finger/silicone paper	Flattening after fixation successful! without losses. Haze formation in the gaps of the surface only
1-4	1x CaLoSiL® E25	flaking is levelled spontaneously because of capillarity. No haze
1-5	2x CaLoSiL® E25	Fine flakes levelled spontaneously; thick scales need flattening
1-6	1x CaLoSiL® E25, flattening with finger/ silicone paper	Flattening directly after the first application successful
1-7	1x CaLoSiL® E25 from behind, flattening after a second application 24h later	Even thick, large scales can be planed and stabilised, but are cracking into pieces
1-8	1x CaLoSiL® E25	Flakes levelled,spontaneously
1-9, 1-10	1x CaLoSiL® E25 grey	Flakes levelled,spontaneously

Figure 7.138 Raking light photos of trial area 1-7 before treatment (left) and after treatment (right).

Figure 7.139 Salone Nero, southern wall, trial areas.

Tests treatments on paint layer damages type (2) were carried out in small area on the base zone of the South wall (see Fig. 7.139). CaLoSiL® was applied under the loose paint layer parts with a syringe. Similarly to the treatment of the thin delamination, it was possible to plane and reapply the loose layers with finger pressure using silicone paper. The results were very promising (see Fig. 7.140).

Figure 7.140 Raking light photo (detail of treatment area TA2) before treatment (left) and after treatment (right).

7.10.4 Evaluation

After the first consolidation as described above, it had to be determined whether the targeted areas could survive a treatment with desalting compresses or solvent compresses without further damage. The tests were promising (see Table 7.29). As an applied water compress did not cause any new damage, desalting seems possible. The solvent compress only caused detachment of some pigment in the black area. A critical survey of the untreated surface revealed that these pigments had been already mobilised by the acrylic fixation in the past.

Eight months after the treatment tests, the results were evaluated. The fixed paint layers (damage type 1) were still stable in most cases. The consolidation of the red and black paint layers was more effective than that of the blue paint layer. The flattening under additional finger pressure achieved better results than without. This

Table 7.29 After-treatments and result

Area no.	Treatments (41% rel. humidity)	Results
1-3	Desalt compress (dist. water)	No damages in the treated area
1-8	Solvent compress (ShellsolT + Acetone)	Mobilisation, of some black pigment
1-9	Desalt compress (dist. water)	No damages in the treated area
1-10	Solvent compress (ShellsolT + Acetone)	No damages in the treated area

Table 7.30 Colour change of the test areas after treatment (gloss included)

Test area	Colour	Treatment	ΔL^*	Δa^*	Δb^*	ΔE^*
1-4	Red	CaLoSiL®	2.11	−0.96	−1.14	2.5
1-7	Black	CaLoSiL®	−11.92	0.53	−0.19	11.9
1-10	Red	CaLoSiL® + solvent compress	5.74	−1.27	−2.18	6.27
2-1	Red	CaLoSiL®	2.11	−0.96	1.14	2.58
3-1	Red	CaLoSiL® + solvent compress	5.74	−1.27	−2.18	6.27
5-4	Black	CaLoSiL®	−11.92	0.53	−0.19	11.93

difference was not detected shortly after the treatment. Pure ethanol had no fixing effect on loose upper paint layer. This clearly showed the strengthening effect of CaLoSiL® in the other test areas. The treated areas of damage type (2) showed no new deformation of the paint layer flakes. The treated impasto paint layer was still stable but new delamination was found nearby. Most test fields showed slight darkening of the colour or dark rims around the treated area. White haze formation was found in the areas of solvent compress tests after the treatment and in a small test area in the base part.

Colour measurements (ISO 11664-4:2008(E), CIE colourimetry-Part 4: 1976 L*a*b Colour Space) were performed to indicate changes to the paint colour (see Table 7.30). Results are given for gloss inclusion; exclusion of gloss gave similar results. In all test areas considerable changes were measured. The changes are mainly observed by an alteration in lightness (ΔL^*).

In the red areas (1-4, 1-10, 2-1) the brightening (increase of L*) is mainly caused by white haze formation . The effect of white haze is more noticeable after treatment with compresses (areas 1-10, 3-1). At the edges of the treated black areas (1-7, 5-4), a very strong darkening was detected. The dark rim of the treatment areas were caused by the dissolution of substances (e.g., acrylic polymer) by CaLoSiL®.

New salt efflorescence was found in the blue area (1-2, 1-3) and in the base area (2-1) where efflorescence had occurred already prior to any treatment. This is an indication of the unchanged porous structure after treatment.

Figure 7.141 Trial area 1-8 (test with solvent compress): new efflorescence 8 moths after treatment.

7.10.5 Conclusion

The evaluation of the trial tests with CaLoSiL® nanosol shows the potential of this material for the pre-strengthening of the damaged paint layers found in Salone Nero (see Table 7.31). The encouraging signs observed directly after the application of the treatments were confirmed by further evaluation after eight months. Additionally the possibility of white haze and the limit of the strengthening effect in case of stronger defects were determined. No indication for densification of the surface by use of CaLoSiL® was found. The

Table 7.31 Compilation of the survey results

Area no.	Haze	Efflorescence	Darkening	Comments
1-1	No	No	Partly	Partly loss of the upper layer
1-2	No	Yes	Rim	New flaking (3–4 mm^2) above white salt crystals
1-3	No	Partly	Slight	Upper part of area: without new deformation; lower part: efflorescence and very small losses in the compress test field
1-4	No	No	Slight	Levelled paint layer still stable
1-5	Partly	No	Rim	Levelled paint layer largely stable, slight haze in position of losses of the upper paint layer
1-6	No	No	Slight	Levelled paint layer largely stable
1-7	No	No	Slight	Levelled paint layer largely stable, slight darkening in the lower part only
1-8	Yes	Yes	Heavy	Levelled paint layer largely stable, some new delamination above salt crystals
1-9	No	No	Heavy	Levelled paint layer largely stable, some new delamination
1-10	Yes	No	Rim	Levelled paint layer still stable
2-1	Slight	Strong	No	Levelled paint layer largely stable, some very small new flaking, strong new efflorescence
2-2	No	No	No	Levelled paint layer partly stable, partly new flaking

Figure 7.142   Trial area 1-10 (test with solvent compress): white haze, dark rim 8 months after treatment.

Figure 7.143 St. Florian statue. Colour marking of areas with different CaLoSiL® E25 consumption.

darkening of the treated areas and dark rims might be caused by the mobilisation of old polymer consolidant by the ethanol solvent of CaLoSiL®.

7.11 Statue of Saint Florian*

The consolidation treatment with $Ca(OH)_2$ nanoparticles in alcohol (CaLoSiL®) was tested on the deteriorated, porous limestone sculpture of St. Florian, located in the village Vratěnin close to house 29 (Czech Republic, region Vysocina/Highlands/). This late Baroque statue (1740) made from Laitha limestone was being restored during 2014 by the restorer Daniel Chadim (Stavebni hut Slavonice Ltd.), who collaborated with ITAM-CET in terms of scientific investigations and consultations.

First, the condition of the sculpture was photographically documented and deterioration patterns recorded and described. Degradation features such as differential erosion (selective elimination of the stone matrix), loss of original surface in some

---

*This section is written by Dana Macounová and Zuzana Slížková.

parts of the statue, biological colonisation, and cracks patterns were predominantly found. After a basic cleaning process including microorganism removal from the stone surface and subsequent drying of the statue, the following stone material characteristics were determined: mineralogical composition and microstructure by optical and SEM microscopy, textural properties by MIP, surface cohesion by peeling test, resistance to drilling by DRMS and water and CaLoSiL® absorption rate by Karsten tube. The second three characteristics were detected in situ, by assessing the condition of selected testing surface areas. The differences between conditions of certain statue surface areas and the changes of characteristics in relation to the consolidating treatment (before and after consolidation) were focused on during this study.

On the basis of stone characteristics (especially pore size distribution and chemical-mineralogical composition) and also the properties of CaLoSiL® consolidants (penetration ability, calcium hydroxide concentration, particles size) the products CaLoSiL® E25 and CaLoSiL® E50 were chosen for structural consolidation of eroded areas and for consolidation of the surface by means of grouting and filling cracks and fissures. First, the consolidant CaLoSiL® E25 was applied on selected deteriorated areas, using squeezy bottle with a glass tube or syringes with needles. Ambient conditions (during and after the treatment) were as follows: temperature 10°C–15°C and relative humidity 70%–80% (the treatment was performed in a restorer's interior studio, where the statue was conditioned for about two months). CaLoSiL® E25 was applied continuously until the consolidated area has been saturated and no more liquid could be soaked into the stone. Any excess of the applied consolidant, especially leaking CaLoSiL® (no infiltrating the stone surface) was immediately absorbed by cellulose or other absorptive materials.

During the treatment, the consumption of the consolidant was monitored on particular surfaces. To enable this, the surface of the sculpture had been divided into several parts, in which their surface area was measured. Thus, the consumption could be expressed in L/m^2 in relation to the given area (see Table 7.33 and Fig. 7.143).

To structurally consolidate the particular parts of the St. Florian sculpture, the area of which is $4.3\,m^2$, $6.5\,L$ of CaLoSiL®

Table 7.32 Scheme of the application of CaLoSiL® products

Treatment	Date of application	CaLoSiL® E25	CaLoSiL® E50	CaLoSiL® paste like
Structural consolidation 1. Treatment cycle	18.02.2014	2 Applications (with about 1 h break)		
Structural consolidation 2. Treatment cycle	20.02.2014	1 Application locally		
Fissures filling (grouting)	20.02.2014		1 Application	1 Application

Table 7.33 The real consolidant consumption expressed in L/m²

Area treated	Area [m²]	Real consumption [L]	Consumption [L/m²]	No. of application cycles
Head, chest, hand	0.60	1.05	1.8	3
Trunk	0.16	0.33	2.1	3
Pennant	0.58	1.25	2.2	3
Skirt, pail	0.81	0.85	1.0	3
Legs, base – front and side walls	0.90	1.30	1.4	3
Rear surface of a sculpture, rear side of the base	0.85	1.25	1.5	2
Fragments of support	0.34	0.45	1.3	2
TOTAL	4.25	6.48	1.6*	3

*average consumption

E25 was used. It took two people three working days to apply the consolidant (including the injection of fractures with more concentrated CaLoSiL® products); see Table 7.32. The surface of the sculpture was treated in two cycles, each comprising a minimum of two applications. In the third cycle, the deteriorated and non-cohesive sites were locally treated. The CaLoSiL® E25 consumption for structural consolidation of the sculpture was 1–2.2 L/m² (depending on the character and damage of the surface). The average consumption was 1.6 L/m².

The application of the consolidant onto the chosen surface during one day of treatment is considered as one treatment cycle, though it actually consisted of more steps. During one day, different selected areas were treated, which were continually saturated with CaLoSiL® E25, so that the solvent did not evaporate entirely and so the surface was pre-wetted for the next application. After each day of application, the stone was covered with polyethylene foil to limit the solvent evaporation rate and to prevent the reverse-migration of the consolidant (it also contributes to the elimination of the white haze). During the first treatment cycle (first day of treatment), the consolidant was applied in two application cycles, one followed by another after about 1 h. The second treatment cycle (second day of treatment) followed three days after the first cycle and within this cycle the consolidant was only applied locally, onto more damaged sites.

In conjunction with the second treatment cycle, filling and grouting fissures and strongly eroded parts of the stone was carried out. These areas were first thoroughly saturated with the more concentrated CaLoSiL® E50 (50 g/L), followed by the application of CaLoSiL® paste like (250 g/L) into cracks and fissures. The amount of CaLoSiL® E50 and CaLoSiL® paste like used for the whole sculpture was about 500 mL and 200 mL, respectively.

After the consolidation, white haze was formed on the surface of the monument. Upon the consulting with the conservator, this layer was subsequently removed by the micro-abrasive cleaning device.

The effect of the consolidation was evaluated using the adhesive tape peeling method. Its principle is the comparison of the weight of the particles removed with the adhesive tape before and after the treatment [150]. The test showed the improved cohesion and

Figure 7.144   Leuben Castle, eastern facade (2008).

consolidation of the stone. In addition to the consolidation effect, the change in the water absorption rate by the limestone before and after the treatment was assessed. The absorption coefficient was measured using the Karsten tube. A decrease in water absorption coefficient was found after the nanolime treatment, which is in accordance with the criteria for positive evaluation of consolidants for porous building materials [27].

7.12 Leuben Castle*

7.12.1 Basic Description of the Object

Leuben Castle is dated from the 17th century and of pure Saxon baroque style. The two-story rectangular building (area c. 42 m × 16 m) has regularly designed facades with window axes and plaster strips. A stuccoed pediment is situated at the eastern facade. The castle deteriorated after 1945 and has been abandoned since 1974, being left to decay (see Fig. 7.144). Whereas a new roof was installed in 2005, the facades and rooms of the castle are currently in a ruinous condition. In recent years, a non-profit association has taken

---

*This section is written by Arnulf Dähne and Christoph Herm.

responsibility for the preservation of the castle. A proposal has been put forward by the Saxon Office for Monument Preservation of Leuben Castle for trial testing of the application of CaLoSiL®.

7.12.2 Test Fields

Research subjects were the stuccoed facade of the eastern pediment (test field 1) and an interior wall with fragments of painting (ground floor, room 012, western wall: test field 2).

7.12.2.1 Test field 1: stuccoed facade (cf. Fig. 7.145)

The pediment is made from mixed masonry (mainly bricks) and sandstone framing. It is coated with two plaster layers from dolomitic lime mortar. The stucco decoration is a free application of fine dolomitic lime mortar above a base layer of coarse stucco and is reinforced with both nails and wires. The coating is lime wash.

The condition of test field 1 was characterised by areas showing loss of both plaster and stucco, together with extensive loss of paint layers. Stucco, plaster layers and visible bricks show heavy damage. The stucco is cracked and fissured in many places. Preserved

Figure 7.145 Leuben Castle, tympanum of the eastern facade, test field 1 (condition prior to tests 2009).

fragments of the paint layers show extensive exfoliation. The fine stucco layer is delaminated. The plaster layers and bricks that are visible show considerable disintegration caused by the crystallisation of salts. The main requirement of the conservation process was to stop the actual decay. Essentially the main aims of the project were to gain the structural strengthening of the coarse stucco while treatment of the fine stucco focused on the grouting of fissures, the repair of loose and delaminated parts, completion of losses, and edging of the fragments.

7.12.2.2 Test field 2: interior wall (cf. Fig. 7.146)

Treatment area TA2 has a flat surface without openings. The mural construction consists of mixed masonry of rubble stones and bricks with dolomitic lime mortar. The wall is coated with two layers of plaster made from dolomitic lime mortars. The first decoration layer is an illusionistic architecture painting, which was probably created using lime paint (however, this was not analysed). This room much like the whole building had been left to decay for decades, which has resulted in the destruction and loss of the floor, the wall

Figure 7.146 Leuben Castle, room 012, bottom zone of the middle part of the western wall, test field 2 (condition prior to tests 2009).

panels and door frames. Plaster and coatings show heavy decay. The plaster has been lost around the test area and its lower parts. The paint layer shows delamination and bulging to different degrees, and there is structural decay of the mortar underneath. Damage patterns observed on the plaster include sanding, flaking, blistering and a weakened structure. As already mentioned, increase of material strength was a specific goal of plaster consolidation. Detachments of the paint were to be reconnected. The necessary treatments covered structural strengthening of the plaster, grouting of plaster detachments, edging of scales, and the repair of the plaster. Flaking paint layers were to be flattened and fixed. Paint layer losses were to be completed and fissures were to be grouted.

7.12.3 Preliminary Tests

Based on the results of the preliminary laboratory investigation, a selection of different application methods were chosen for the trial tests. Initially, treatments using nanosols without pre- or post-modification were used. Different concentrations of nanosol were also tested. The addition of acetone and some pre- and post-treatments were applied in order to enhance depth distribution and to avoid white haze formation. As post-treatment with pure solvents did not affect the white haze, hydroxypropyl cellulose (HPC) gel, both as an additive to the consolidant and as a post-treatment, was tested to delay solvent evaporation. For all applications on the test area pipettes were used. In the cases of repeated application, the interval between two single applications was at least one week.

Strengthening with unmodified CaLoSiL® (25 g/L) produced a white haze on the surface after only a single application. The effect appeared to increase with raised levels of moisture in the substrate material. The white haze formation could be reduced by dilution of CaLoSiL® or addition of a HPC gel of low concentration. post-treatments with HPC solutions proved more effective in reducing white haze than pure ethanol.

Better results were observed by using a combination of treatments rather than single methods. The most promising treatments were with 50% diluted CaLoSiL® NP25 and the application of CaLoSiL® type E (including acetone) and the stepwise increase in concentration during the single applications, combined with post-treatment with HPC-solution. The HPC post-treatment seems to be essential to avoid the remigration of $Ca(OH)_2$ onto the surface of the substrate. In all cases the resulting strength was not fit for purpose and further studies were needed.

The poor strengthening effect of nanosol during previous testing seems primarily due to the insufficient distribution of the strengthener in the substrate. The effects of modifying the CaLoSiL® formulation and evaluating different application methods were investigated by measuring depth distribution. This second phase of laboratory work also involved tests on the modification of CaLoSiL® for use as a grout and putty. The results are documented above.

7.12.4 Trial Conservation

7.12.4.1 Test field 1 (see Fig. 7.147)

For structural strengthening of the coarse stucco, an effective modification of CaLoSiL® was to be identified on the basis of the laboratory test series. The strengthener should consolidate sanding and crumbling surfaces and increase the strength of the weak material in general. It was essential to avoid the formation of white bloom on the substrate. After preliminary tests on the object, a mixture of CaLoSiL® NP50 + CaLoSiL® micro + acetone $(200 + 1 + 50, m/m/m)$ showed the best results and was applied to all surfaces of the coarse stucco in the demonstration area using a pipette until saturation. A small area was tested using an immediate coating of 0.5 m% HPC gel in n-propanol (single application with pipette) to the substrate surface. The fine stucco was found to have only minor detachments and fissures. These were to be consolidated with modified CaLoSiL® depending on their dimensions, following the approach based on the laboratory test series. Any endangered edges and all losses within the preserved layer were puttied with a material based on lime dispersion. Gaps and fissures were filled by injection with syringes using thickened nanosol (10 volume parts CaLoSiL® E50 + 5 volume parts CaLoSiL® micro + 1 volume part HPC gel (0.5 m% in ethanol/water 75 vol%)). Depending on the situation, three to four application steps were necessary. The

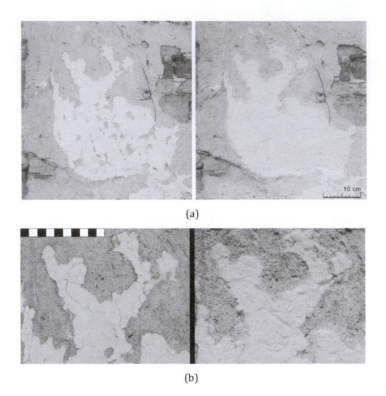

(a)

(b)

Figure 7.147 (a) Leuben Castle, eastern facade, Tympanum, left part: test field 1 before (left) and after (right) the treatments. (b) Leuben Castle, eastern facade, tympanum, left part: Detail of test field 1 before (left) and after (right) the treatments.

application of putty to any losses and edging of the fine stucco were performed with a paste made from CaLoSiL® micro and lime chalk (chalk of Champagne) (36 + 64, m/m). Colour adjustment of the putty was not necessary. The substrate was pre-wetted with ethanol prior to the application of the putty.

7.12.4.2 Test field 2 (see Fig. 7.148)

To strengthen the structure of the plaster, laboratory testing was used to identify an effective modification of CaLoSiL®. The strengthener should consolidate sanding surfaces and increase the

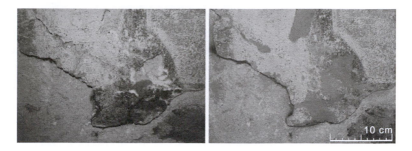

Figure 7.148 Leuben Castle, room 012, eastern wall, middle part, bottom zone: Detail of the test field 2 during (left) and after (right) the treatments. Losses in the lower part were caused by the drilling resistance measurement.

strength of the weak material in general. As mentioned previously it was essential to avoid the formation of white bloom. Gaps of plaster detachments were present in dimensions of about 2–10 mm. These areas were to be edged with putty and grouted with an injection mortar on the basis of results from the laboratory tests. During preliminary testing to investigate structural strengthening of the plaster two methods showed positive results. Both were applied in different areas:

- CaLoSiL® NP50 + CaLoSiL® micro $(100 + 1 m/m)$/application with pipette until saturation; followed by immediate post-treatment with 0.5 m% HPC in n-propanol (single application covering of the surfaces with pipette)
- CaLoSiL® IP50 + CaLoSiL® micro $(200 + 1 m/m)$ + acetone $(4 + 1 m/m)$/application with pipette until saturation

Both treatments were repeated twice, observing at least one week between applications. The edges of detached plaster were puttied previously with a mixture of CaLoSiL® micro + lime chalk (chalk of Champagne) + mixture of earth pigments $(4.5 + 6 + 2 m/m/m)$ after pre-wetting with ethanol. The grout was made from CaLoSiL® E50 + CaLoSiL® micro + filler mixture (glass microspheres and chalk of Champagne) $(6 + 6 + 1 m/m/m)$. The application with syringes was repeated until the void was completely filled.

Plaster completion was tested in several larger and deeper losses within well preserved paint layer areas. A repair mortar was made from CaLoSiL® micro, quartz sand (grain size 0.04–1.00 mm) and earth pigment mixture (2.5 + 10 + 0.4 $m/m/m$) and the application was performed after pre-wetting with ethanol. The completed surface was directly below the level of the paint layer.

Fissures were filled by injection with syringes using thickened nanosol: [CaLoSiL® E50 + CaLoSiL® micro (2 + 1V/V)] + [0.5 m% HPC in ethanol + water (3 + 1V/V)] (5 + 1V/V). Depending on the case three to six applications were necessary.

Following the results obtained in the test series, any paint that had become loosened from the surface was to be fixed with modified CaLoSiL® dependent on dimension. The potential to flatten flakes during the consolidation treatment was also tested. Fissures and larger scaled delaminations were to be filled with grout.

All deeper openings (losses and fissures) were filled with thickened nanosol using syringes: [CaLoSiL® E50 + CaLoSiL® micro (2 + 1V/V)] + [0.5 m% HPC in ethanol + water (3 + 1 V/V)] (5 + 1 V/V). Four to six measures were necessary. The flattening of flakes was found to be possible. Tiny flakes flattened themselves by the suction effect of the liquid strengthener. Larger flakes could be flattened during the consolidation treatment by finger pressure (on PE-film as protection layer). Detachments of a smaller dimension were treated with a mixture of CaLoSiL® E50 + CaLoSiL® micro (1 + 1 V/V) using syringes for application.

The application of putty to the paint layer was tested on several smaller losses and fissures. These completions should be made with the same level and surface smoothness as the surrounding preserved paint layer but with a neutral colour. The used putty was made from CaLoSiL® micro, lime chalk (chalk of Champagne) and earth pigment mixture (4.5 + 6 + 2 $m/m/m$). Pre-wetting with ethanol was carried out before application.

7.12.5 Evaluation

7.12.5.1 Test field 1

The surface of the coarse stucco showed very good consolidation after treatment. Only heavily deteriorated parts could not be stabilised effectively. Furthermore, white bloom formation was found, despite the fact that small pre-test areas had not shown any appearance of bloom earlier. Thus, it is important that results of a general nature should be viewed with caution. Drilling resistance measurements did not produce any results that were useful. The fine stucco appeared sufficiently consolidated and parts that were previously loose were fixed. The stability of the stucco was such that it would withstand further physical treatments, for example mechanical cleaning. Strength measurements were not performed. Edging and completions appeared stable without any visible shrinkage (cf. Fig. 7.147). However, all surfaces were powdering.

7.12.5.2 Test field 2

The consolidation of the plaster surfaces and structural strengthening were successful to a certain extent. Sanding of surfaces was remedied almost completely. The previously weak material showed increased stability against the physical effects of further conservation treatments. However, the achieved strength was still not strong enough to act as a stable substrate for repair mortar. Furthermore, white bloom occurred, although small dimensioned pre-tests on small areas had not shown any. The fixing of detached plaster by edging seemed strong enough to prevent injected mortar from flowing out after application; however powdering was observed at the surfaces. Seemingly, this material does not appear to be durable in the long term. Gaps and voids were completely filled after several applications of the grout. Manual palpation of the detached plaster was sufficient for its stabilisation. The grouting however showed very poor strength on analysis of drilling resistance measurements. Further tests should be done to evaluate the performance of the grouts. Similarly, completions of plaster losses showed very weak consistency. On a more positive note, fissures could be completely filled after several applications of the grout. Measuring any improved stabilisation was not possible.

The selective treatment of areas which showed different degrees of paint layer decay resulted in consolidation of the original surface that was acutely damaged. The fragile plaster below the damaged paint layer could also be consolidated by treatment (cf. Fig. 7.149).

Figure 7.149 Leuben Castle, room 012, eastern wall, middle part: Detail of test field 2 before (left), during (middle) and after (right) the treatments.

One major problem that occurred during the application of consolidants was the difficulty to determine the adequate amount of consolidant required each time. As consequence of this a rather high number of applications were necessary to achieve acceptable results on all test areas. Another difficulty is the unavoidable contamination of the paint with the nanosol during the treatment resulting in white stains appearing on the surface of the substrate. Immediate cleaning with solvent is relatively straightforward and often successful but is impossible in circumstances where areas of the paint layers are still fragile. In these situations, the removal is only possible, but still not easy, after effective consolidation. As any removal of surplus consolidant was only possible using acetic acid, consequential risks have to be considered. The brittle character of the binding after consolidation implies sensitivity of the treated paint layer to the mechanical stress that is inevitable during post-consolidation activities such as cleaning or the application of putty. The repairs carried out on paint layer losses appeared stable without any visible shrinkage; however, again all surfaces were powdering.

7.13 Dahlen Castle*

7.13.1 Basic Description of the Object

Dahlen Castle (see Fig. 7.150) was built between 1744 and 1751 and is a two-story building complex based on a H-shaped ground plan (ext. dimension ca. $40 \times 40\,\mathrm{m^2}$). It has regularly designed facades

---

*This section is written by Arnulf Dähne and Christoph Herm.

Figure 7.150   Dahlen Castle, west side (2008).

and window axes and its rooms were embellished in a baroque style by Christian Oeser. In 1974 the castle was badly damaged by a fire and as the original roof has not been preserved, the building is now covered by a roof added as an emergency solution. The rooms of the castle are currently in a ruinous condition and for the last few years, the preservation of the castle has been overseen by a non-profit association.

7.13.2 Test Field

A proposal by the Saxon Office for Monument Preservation was put forward to select the 'White Saloon' of Dahlen castle (see Fig. 7.151) as a candidate for the trial testing of CaLoSiL® as a consolidation agent. This rectangular room (sized 8.55 m × 8.55 m × 4.50 m) is situated in the middle axis of the ground floor at the eastern side of the castle. Since Baroque times it has been completely decorated with stucco panelling and a painted ceiling and is considered to be one of the most magnificent rooms of the castle. Today, however, it is

Figure 7.151 Dahlen Castle, 'White Saloon' corner area of western and northern walls (2008).

in a ruinous state. While fragments of the former stucco decoration are preserved, the original ceiling has not survived. The northern part of the western wall was chosen for trial tests (Fig. 7.152).

The mural construction consists of mixed masonry of rubble, stone and bricks. The ground plaster is made from gypsum mortar which is covered by several layers of white gypsum stucco with iron reinforcements. Only fragments of the white gypsum stucco have been preserved. Very few parts of the original surface have remained with stucco material and ground plaster being heavily damaged by the infiltration of rainwater. The stucco shows multi-layered scaling and delamination, weak cohesion, and many parts have broken away. The iron reinforcements are rusty and the cohesion of the ground plaster has been lost resulting in an extremely friable consistency. The plaster only exists in thin layers and all open surfaces are sanding.

The main conservation requirement was to stop the acute decay of both the plaster and stucco. Consolidation was to be used to

Figure 7.152 Dahlen Castle, 'White Saloon' western wall, northern part with the test field (condition prior to tests 2009).

increase the structural strength of the any material that showed deterioration. The following specific treatments were required: structural strengthening, the fixation of any loose parts, scales and detachments as well as the edging of scales and damaged edges.

7.13.3 Preliminary Tests

Initially, for the treatment of gypsum bound materials, the pilot product $CaSO_4$-nanosol was chosen because of its similarity in composition to the test substrate. Trial tests on the deteriorated gypsum plaster and stucco with $CaSO_4$-nanosol were only successful for the stabilisation of superficial damage like surface powdering. Little success was achieved in the structural strengthening of gypsum plaster and stucco under given field test conditions. The underlying reason for this was that depth distribution was insufficient because of the influence of the moisture content of the material. As the $CaSO_4$-nanosol in its current form had proved ineffective, it was decided that further application of this product was not the most worthwhile option.

For further tests, CaLoSiL® calcium hydroxide nanosol was chosen. Again, test treatments of deteriorated gypsum plaster and stucco marble with CaLoSiL® E25 were only successful in the stabilisation of superficial damage like surface powdering. The effect on the structural strength was observed to be better on gypsum mortar with a low moisture content. The increase in stability recorded, however, was low even after six applications. In both test cases the formation of white haze was noted. The treatment tests carried out on the white stucco showed varying results. On one hand, flaking white stucco with a low moisture content was not stabilised by six applications of unmodified CaLoSiL® E25; moreover the structure even appeared to soften to some degree after the treatments. Conversely, the strengthening of weak, powdering stucco with a high moisture content was successful. Furthermore, the moisture content of the material seems to have an influence on the depth distribution of CaLoSiL® E25 and any subsequent formation of white haze.

The application of CaLoSiL® E with acetone as an additive, and a stepwise increase in concentration during single applications was found to stabilise the flaking of white stucco and the heavily deteriorated gypsum mortar to some extent without the formation of white haze. The increase of structural strength achieved might be sufficient as a pre-strengthening treatment for any subsequent conservation work, but it was still moderate at best. Preliminary tests using modified CaLoSiL® nanosols with bi-modal particle size distribution (see Section 5.2.2) showed promising results. The best results, at least for the strengthening of white stucco, were achieved using a mixture of CaLoSiL® E25 + CaLoSiL® micro ($100 + 1 \, m/m$).

7.13.4 Trial Conservation

An application of a mixture CaLoSiL® E25 + CaLoSiL® micro ($100 + 1$, m/m) was used to strengthen the structure of white

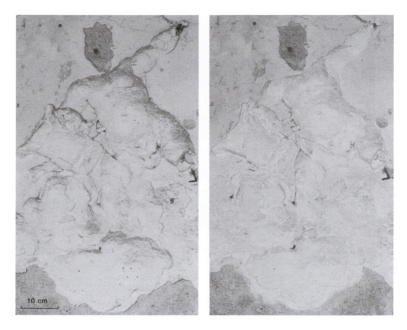

10 cm

Figure 7.153 Dahlen Castle, 'White Saloon' western wall, right part (part of UFAD-2-TA1): Demonstration area before (left) and after (right) the treatments.

stucco. The mixture was applied using a pipette and this process repeated until the whole area was saturated. The treatment was then repeated twice with a weeks break between applications. Gaps of detachments and scales as well as fissures and cracks were filled by the injection of thickened nanosol (10 volume parts CaLoSiL® E50 + 5 volume parts CaLoSiL® micro + 3 volume parts HPC gel 0.5%m in ethanol/water (3 + 1 *V/V*). A triple application was necessary for sufficient consolidation. For the puttying of scales and edges, a paste was made from CaLoSiL® micro, inert gypsum powder (chalk of Bologne) and earth pigment (1 + 2 + 0.01 *m/m/m*). The application was performed after pre-wetting with ethanol. Ground plaster was not included in the demonstration area because no adequate strengthening method was identified for its successful treatment in previous tests.

7.13.5 Evaluation

In summary, white gypsum stucco showed increased structural strength at its surface after the treatments. The material did not show powdering and is stable enough for mechanical cleaning. The consolidation of all deteriorated decorations such as loose parts, scales, and detachments appeared to be successful as tested by palpation. The putties used for edging were stable without any visible shrinkage; however, they did show powdering at their surfaces.

7.14 Consolidation of Paint Layer with Lime Nanosols*

7.14.1 Introduction

The consolidation of wall painting and architectural surfaces is connected with the idea of compatibility, as is every other field of monument heritage care. A commonly accepted opinion is that a consolidated material has to be treated with a material which is similar in terms of both its physical properties and chemical composition. Therefore, the use of lime nanosols (or nanosuspensions) for the consolidation of paint layers on plaster, often also bound with lime, would appear to be ideal in many ways. As demonstrated in several studies [76, 82, 151], the use of lime water to date, which is also based on consolidation with calcium hydroxide, has proven to be extremely time-consuming and problematic in other aspects. The highest risk of lime water application lies in the very low concentration of an active substance and the necessity of its application in several cycles, resulting in a considerable moistening of the object, thus activating degradation processes such as the effects of water soluble salts or frost cycles, etc. During the STONECORE project, it became obvious that lime nanosols could normally be used to consolidate the paint layer in concentrations of 5 to 10 g/L, which is roughly 3 to 6 times higher than in the case of lime water. The fact that alcohol is a carrier of calcium hydroxide in lime nanosols is, of course, a further advantage

---

*This section is written by Jan Vojtechovsky.

compared to lime water because the reinforced material is not moistened and the time between application cycles is also reduced due to the faster evaporation of alcohols.

7.14.2 Trials on Test Panels

Although some publications related to the use of lime nanosols for the consolidation of paint layers in wall paintings existed before the project [152], they did not contain much factual information. This is why we basically had to start from scratch with the tests [153].

To begin with, we had to carry out tests on sample panels that simulated real paint layers. We thus prepared test panels consisting of a wood wool-cement board, onto which weakly-bonded lime plaster was applied in two layers (coarser and finer), and on top of which the wash of simulated paint layer was then applied. The simulated paint layer, which contained no binder, consisted of a mixture of clay pigment (burnt umber), and limestone powder. The panel was then divided into test areas of 10×10 cm.

7.14.2.1 Selection of a suitable consolidant

Within the scope of the project, we tested several suspension formulae with different concentrations and different dispersion media. For this purpose, various alcohols were tested and observed with respect to their influence on the behaviour of the resulting suspension. The most noticeable behaviour that we were able to observe in the tests was the extent and type of veil that formed after the application. Consequently, we came to the relatively explicit conclusion that the best results were achieved with suspensions for which ethanol was used as a dispersing medium. Other aspects of the veil formation such as the concentration of the active substance, method of application and the number of cycles will be discussed below.

7.14.2.2 Method of application

The risk of creating a white haze on the surface of the object being consolidated is one of the main problems of the application of lime nanosols. However, this is a crucial aspect when it comes to wall paintings or architectural surfaces because the goal of an ideal method of application is, logically, a treatment that ensures the required consolidation without any aesthetic change in the material being consolidated.

Until recently, applying the suspension with a brush over Japanese tissue [152] has been the recommended method of application. However, this method did not work on the panels on account of the almost immediate appearance of a strong white haze. There was thus an urgent need for other application procedures. After several tests, spraying proved itself to be the best application method. To ensure a uniform application of spray, it is obviously preferable that the suspension aerosol be as fine as possible, because only then can an even result be achieved. However, some sprayers were unable to perform the task, leaving larger or smaller drops after the application, which meant an uneven distribution of consolidant and a change in aesthetics. Ultimately, a sprayer with a gas cartridge proved to be the best option. Nevertheless, for certain wall paintings, especially those suffering from serious deterioration, application with a brush over Japanese tissue, or even better with a syringe, may not be that risky and is more effective than spraying (due to the larger amount of material that can be applied in this way).

It is the amount of the consolidant that determines the ultimate result of a consolidation. Consequently, the goal of our work was to find the most suitable setting for a consolidant with the best parameters in terms of concentration, number of applications and method of application so as to achieve the desired results. This means we were looking for the highest possible concentration and the lowest possible number of cycles to ensure a sufficient consolidation, without any significant aesthetic changes. During the tests on panels, it was discovered that the maximum possible concentration of the suspension for this type of substrate is 10 g/L. However, this concentration can only be applied in three cycles without causing any significant problems (the formation of white haze), whereas applying the suspension at a concentration of 5 g/L proved more favourable in terms of its control since it could then be applied in up to 8 cycles without producing any noticeable haze. Nevertheless, certain rules have to be followed so as to minimise the risk of the formation of a white veil. These are mainly:

- The climatic conditions are the primary factor for the application of the suspension, in no matter how this is carried out. The lower the relative humidity, the higher the risk of the formation of white haze. Although it is difficult to prove the exact mechanism related to this aspect, the most likely explanation is a combination of the back-migration of solid particles during evaporation of the solvent and the formation of a specific, crystalline calcium carbonate structure in low humidity conditions. It is therefore advisable to apply a suspension when the relative humidity is around 60% and above.

- The fact that the amount of consolidation material significantly affects the formation of white haze must be born in mind with any method of application. As soon as the suspension shows the first signs of its deposition on the surface of the material being consolidated, the application procedure must be stopped. In order to get a sufficient amount of reinforcing material into the material being consolidated, it is necessary to increase the number of cycles (repeat the procedure).

- Several experiments were performed with pre-wetting and subsequent wetting with water and various organic solvents (mostly with alcohols which were also used as a dispersing agent). However, these showed that pre-wetting should not be carried out. It would appear that any closure of the pores of the material being consolidated with any liquid leads to the formation of a strong haze during the subsequent application of the suspension. As for subsequent wetting, the best results were achieved when using water. It appears that water has a positive effect on the reduction of the back-migration of the calcium hydroxide particles to the surface and it may also influence the crystalline structure of calcium carbonate that is being formed. This is the only way we are able to explain the fact that, although the veil had already formed after several cycles of applications, it could be significantly suppressed by subsequent wetting (see Fig. 7.154). To as much as possible reduce the risk of white hazing, water should be applied after each cycle of the suspension application. On the other hand, in this procedure, there is the risk of

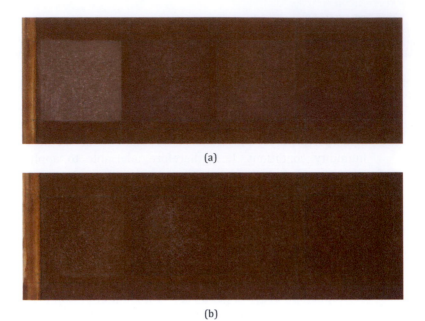

(a)

(b)

Figure 7.154 Trials on test panels: (a) The state after 6 application cycles with various lime suspensions at a concentration of 5 g/L without any subsequent wetting. (b) The state after the subsequent application of 2 cycles of water: the veil was reduced, even though the application was carried out ex post facto.

a darkening effect or even deterioration (microcracks). This generally affects degraded and damaged paint layers and can occur after multiple wetting. The experiment also showed that wetting after each application cycle of the suspension has a negative effect on the resulting strength of the substrate. It would therefore be preferable if the substrate being reinforced allows the possibility of applying water after every second or third cycle, for example.

- Nevertheless, the reinforced material itself is probably the most important factor that determines whether a white veil eventually forms or not. The pore size of the consolidated substrate naturally plays a crucial role. However, the deterioration of the material, which is a basic condition upon which the procedure of consolidation can be performed, is very important too. The more open the structure of the consolidated material (either because of its original properties, or as a result of its poorer state), the greater the amount of consolidating agent that can penetrate the material without forming a white veil. Tests showed that in the case of extremely degraded materials, the risk of white haze formation is relatively low, but at the same time the consolidation may be insufficient, even if a relatively high number of cycles is performed. If the haze formation is not a real threat, however, a high number of cycles should not pose a problem. In the tests, however, it could be shown that in the event of a high number of cycles, the surface pores are filled and closed before any satisfactory filling is achieved below the surface. The result is that a compact white crust already starts to form on the surface, whereas the consolidation is still insufficient below the surface.

7.14.2.3 Tests in situ

After extensive tests on model panels and a pre-selection of the most suitable suspensions and methods of application, testing in situ began to validate the achieved results. The STONECORE project included tests of paint layer consolidation, which were performed at two different sites by the Faculty of Restoration, University of Pardubice. These were in the Church of Saint Vitus in the abandoned village of Zahradka, which is now located on the shore of the Svihov reservoir about 18 km northwest of Humpolec, and in the Chapel of Saint Isidor in Krenov, which is located about 11 km south of Moravska Trebova. Whereas with the wall paintings in Zahradka (see Fig. 7.156a), there was moderately powdered medieval painting, which occurred on a relatively compact plaster, the ceiling frescoes in the Chapel in Krenov (see Fig. 7.157a) were extremely degraded paintings and plasters, which had been damaged due to long-term exposure to leaking rainwater. That was one of the reasons why different approaches were chosen for these two specific paintings.

The ethanolic nanolime dispersion was selected for the paintings in Zahradka (based on trials) and applied in two phases. Firstly, pre-consolidation was carried out in two cycles at a concentration of 10 g/L. Following cleaning of the painting, two more cycles of the suspension were applied at a concentration of 5 g/L. In both cases, subsequent wetting with water was carried out to reduce the risk of the formation of white haze. What's more, the relative humidity did not fall in either case below the aforementioned recommended level of 60%. The level of consolidation can be assessed as satisfactory; there were no signs of white haze formation on the painting surface (see Fig. 7.156b).

The consolidation of the paintings on the vault of the Chapel of Saint Isidor in Krenov was a somewhat different case. As mentioned above, very significant damage in the form of strong powdering and blistering of the paint layer had been caused by substantial and persistent leaking of rainwater, related to the earlier inappropriate application of a polymer-based consolidant. In addition, the intonaco layer was separating, resulting in the appearance of large open blisters (see Fig. 7.155). Although the application of the suspension to the painting at a concentration of 10 g/L in roughly 5 cycles (taking into account the degree of disintegration) was completely sufficient in many areas, in some places, the attempt to achieve a satisfactory result failed, even after 12 suspension cycles at the aforementioned concentration. Moreover, the subsequent consolidation with 1% hydroxypropyl cellulose in two cycles appeared to be ineffective.

Therefore, after finishing the STONECORE project, we proceeded by returning to examine the test panels. Since most of the existing damage to the painting was related to the action of moisture and the previous application of the polymer coating, which had a very negative impact on the painting, we wanted to stick to consolidants on an inorganic basis. According to the laboratory tests which were carried out [154, 155], the most appropriate alternative appeared to be a combination or a mixture of nanolime with silicic acid esters. We decided to pursue this approach, not only because this combination had proven effective in the case of other materials [156]. After testing in situ, in which we investigated different ratios and concentrations of the examined consolidants, we selected the final consolidation technology. Firstly, the application of a mixture based on silicic acid esters at a concentration of 300 g/L and a lime-

Figure 7.155 Open blisters of intonaco in the vault of St. Isidor's Chapel in Krenov.

ethanol nanosuspension at a concentration of 10 g/L at a ratio of 1 : 1. Subsequently, the final consolidation was carried out with an ethanol-lime nanosuspension of 5 g/L in 3–6 cycles. In those areas where even this combination was unable to achieve an adequate strengthening, the mixture described above was applied again. In the case of the final technology, the degree of consolidation can be rated as sufficient, since there were no signs of white haze formation on the painting surface (see Fig. 7.157b).

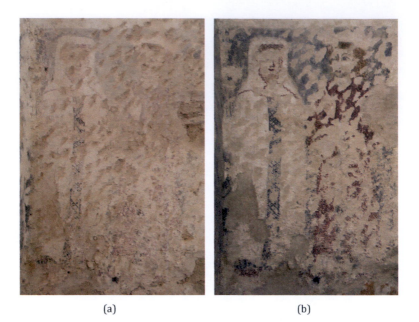

(a)   (b)

Figure 7.156   Part of a wall painting on the south wall of the presbytery of the St. Vitus Church in Zahradka. (a) The state before the treatment. (b) The state after the conservation.

7.14.3 Conclusion

From the tests carried out and from the restoration treatments on the selected objects, we can conclude that the consolidation of lime-based wall paintings and architectural surfaces using lime nanosuspensions is a very promising option for the future that can appropriately complement, or in some cases even completely replace, those technologies currently in use. However, as explained above, this technology harbours certain risks that have to be taken into account. The main, obvious risk is the formation of a white veil during application. Nevertheless, in the foregoing text we have described methods to effectively prevent this. One specific disadvantage that can be named, though only in the case of extremely degraded materials, is the inadequate consolidation ability. One solution that was put to successful use in the restoration treatment of one of the paintings included in the project, could

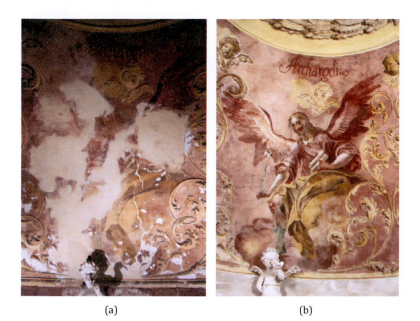

(a)   (b)

Figure 7.157   View of part of the vault of the St. Isidor's Chapel in Krenov.
(a) The state before treatment. The painting is extensively damaged due
to rainwater leakages. (b) The state after treatment with a reconstruction
based on historical photos performed with the water colour technique.

be a mixture or a combination of nanosuspensions and silicic
acid esters. This retains the advantage of a conservation with
inorganic material whilst preserving the compatibility with the
treated material. This solution and its long-term impacts, however,
have yet to be investigated in more detail.

Chapter 8

Nanolime: Recent Publications and Application Recommendations

Gerald Ziegenbalg and Claudia Dietze

IBZ-Salzchemie GmbH & Co. KG, Schwarze Kiefern 4, 09633 Halsbrücke, Germany
gerald.ziegenbalg@ibz-freiberg.de; claudia.dietze@ibz-freiberg.de

8.1 Recent Publications

There is a growing interest in the use of nanolime in different applications that has resulted in an increasing number of publications over the past few years. It is impossible to discuss all of these papers in detail in the context of this book. Many of them have already been cited and discussed in previous chapters. The main aspects will be summarised and compared with own investigations in the following.

Nanolime dispersions have been applied successfully to numerous materials, especially for the consolidation of decayed stone and mortars as well as wall paintings. The following literature constitutes a small selection of some applications concerning nanolime [55, 89, 112, 118, 152, 157–168]. A comprehensive and detailed description of the application of nanolime on special limestones can be found in Refs. [169, 170]. The successful use of

Nanomaterials in Architecture and Art Conservation
Gerald Ziegenbalg, Miloš Drdácký, Claudia Dietze, and Dirk Schuch
Copyright © 2018 Pan Stanford Publishing Pte. Ltd.
ISBN 978-981-4800-26-6 (Hardback), 978-0-429-42875-3 (eBook)
www.panstanford.com

nanolime for the conservation of ornamental hardened limestone is described in Ref. [171]. The use of nanolime for the conservation of weathered limestone is discussed in detail in Ref. [172].

The consolidation effect of nanolime is based on carbonation, resulting in the formation of various calcium carbonate polymorphs such as amorphous calcium carbonate, monohydrocalcite, vaterite, aragonite and calcite. Their formation/transformation depends largely on the curing conditions such as relative humidity and temperature as well as the properties of the substrates. The influence of different environmental conditions is discussed extensively in the following literature [66, 69, 132, 173]. The presence of alkoxides, which may form during storage of the nanolime dispersions, also affects the rate of carbonation and the resulting polymorphs [129]. Despite the successful use of nanolime dispersions as consolidation agents, one main drawback is the possible reverse-migration of these nanoparticles, leading to a white haze and a reduced consolidation zone. The results obtained by Dähne and Herm [133] on the reverse-migration of nanolime during evaporation of the alcohol (see Chapter 4) were confirmed by Borsoi et al. [174], who studied the penetration of CaLoSiL® E25 into Maastricht limestone. Although an excellent penetration had been achieved, the high volatility of ethanol together with the high stability of the dispersion facilitated the partial reverse-migration of nanolime particles. Borsoi et al. state that the trend of the capillarity wetting curves can be related to the sorptivity (S), which is a measure of the capacity of porous material to absorb and transmit liquid by capillarity. The sorptivity is related to the surface tension of the liquid and the dynamic viscosity (η).

$$S = \sqrt{(\sigma/\eta)} \tag{8.1}$$

It has been found that drying Maastrich limestone treated with nanolime has resulted in the formation of an internal drying front, situated approximately 0.5 mm below the treated surface. No white haze deposition has been detected. Different approaches were investigated to avoid reverse-migration. Borsoi et al. [175] found that the best consolidation of the porous Maastricht limestone could be achieved if the nanolime dispersions (prepared by a special solvothermal synthesis) consist of a mixture of ethanol and water at the volume ratio 95% : 5%. In this context, it should be mentioned that CaLoSiL® always contains up to 2 wt% water. In another study by Borsoi et al. [176], relatively coarse nanolime particles (40–800 nm) were solvothermally synthesised and dispersed in ethanol, *iso*-propanol and *n*-butanol, respectively. As expected, the dispersions in butanol displayed the lowest evaporation rate. All of the dispersions were able to penetrate Maastricht limestone. Whereas treatment with the dispersions containing ethanol resulted in a white bloom after evaporation, *n*-butanol-based suspensions did not exhibit any reverse-migration. It should be remembered, however, that the particle size of the dispersed nanolime in *n*-butanol was much higher than in ethanol. Thus, it can be assumed that the reverse-migration was mainly prevented for steric reasons. The same has been found by Herm and Dähne, who demonstrated that adding small amounts of CaLoSiL® micro to CaLoSiL® E25 significantly prevented reverse-migration [133].

The consolidation of Maastricht limestone using nanolime is once again described in Ref. [177]. Based on sophisticated X-ray radiography combined with drilling resistance measurements, the distribution of nanolime in cubes with the dimensions $20 \times 20 \times 40 \ mm^3$ was investigated. The cubes were saturated through immersion in a constant volume of 2 mL CaLoSiL® E25 and E50, respectively. The X-ray radiography clearly identified zones of higher density caused by the inhomogeneous distribution of nanolime due to reverse-migration of the particles during drying. A two-stage treatment process based on nanolime application (2 mL) followed by the application of water (0.5 mL) resulted in a homogeneous distribution of the particles within the cubes. The water was applied directly after saturation of the sample with CaLoSiL®.

That the formation of white haze could be prevented by the combined application of nanolime and water was confirmed by [153]. Both a high relative humidity and subsequent wetting with water restricted the formation of white haze. A relative humidity of around 60% was found to be sufficient. Working at a low RH of 30% has been regarded as a risk. It was also discovered that even when white haze had formed, it could be reduced by wetting with water. However, the white haze that had formed could not be completely eliminated by wetting with water. If several repeated applications of nanolime dispersions are necessary, wetting should be performed once every 2–3 cycles. Wetting after each application cycle has a negative effect on the final strength. A sprayer with a gas cartridge that produces a fine aerosol was found to be the ideal application method for the stabilisation of wall paintings. This ensured a uniform distribution during application.

The consolidation of lime mortars with CaLoSiL®, Nanorestore® and lime nanoparticles from Merck is described in Ref. [178]. Iso-propanol based dispersions were used in concentrations of 5 g/L. A 25 g/L nanolime dispersion of CaLoSiL® has also been tested. Cylindrical specimens based on lime hydrate-sand mixtures (1 : 3 by weight) with a height of 4 cm and diameter of 3 cm were prepared and cured at 20 °C ± 5 °C at 60% ± 5% RH for 28 days. The nanolime dispersions were absorbed into the mortar samples by capillary suction. Dry samples as well as samples saturated with water before the nanolime dispersions were applied have been studied. In addition, several samples were saturated after the application of the nanolime dispersions, again with water. The samples were stored at an RH of 90% ± 5% at 19°C in partially ventilated chambers. The samples were weighed after 8 weeks and oven dried at 30°C and 20% RH for 24 h. In all samples carbonation had been achieved within this time, though to different degrees. Apart from the samples treated with the 25 g/L CaLoSiL® product, calcite as the only calcium carbonate modification found. Surprisingly, aragonite was found as a second $CaCO_3$ modification in the sample treated with the 25 g/L dispersion. The 25 g/L dispersion of CaLoSiL® reached only a low penetration depth in a pre-wetted sample. Based on experiences obtained in the STONECORE project, the authors of this chapter assume that the presence of water resulted in a rapid coagulation of the nanoparticles, leading to a dense layer of calcium hydroxide particles on the bottom of the sample.

One main aspect in Ref. [153] concerns the phase morphology and texture of the resulting calcite crystals. Treating non-water saturated samples with 5 g/L dispersions of CaLoSiL® resulted in a compact matrix composed mainly of rhombohedral and scaleno-hedral calcite crystals. Pre-saturation with water produced smaller and more dispersed calcite crystals as well as aragonite. None of the mortar samples treated with the 5 g/L CaLoSiL® dispersion displayed any significant chromatic variations. It is concluded, 'Of the three different products, CaLoSiL® induced the highest degree of carbonation. When applied at a concentration of 5 g/L, the CaLoSiL® dispersion was a more effective consolidating compound than the Nanorestore® and Merck dispersions, in that it induced the highest degree of carbonation, decreased mortar porosity and made the matrix more compact thanks to the formation of densely aggregated rhombohedral calcite and needle-like aragonite crystals.'

The importance of the environmental conditions for the carbonation of nanolime dispersions (Nanorestore® 5 g/L, CaLoSiL® E25) is summarised in detail in Ref. [179]. It could be shown that higher carbonation is achieved when this takes place in an atmosphere saturated with H_2O, CO_2 and ethanol. The lowest carbonation rate was detected under laboratory conditions at 56% RH, whereas 85% of unreacted portlandite was still present after 28 days. Similar results were obtained in Ref. [173]. Relative humidities between 75% and 90% result in the formation of ACC, MHC, calcite, vaterite and aragonite. Complete carbonation was already achieved after 7 days. At an RH of 33%–54%, large amounts of unreacted portlandite and only 5%–8% of MHC were still present after 28 days. 8% calcite was present after 28 days at an RH of 33% and vaterite formed at RH of 54%.

Detailed investigations into the combination of nanolime dispersions with silicic acid esters have mainly been published by the authors of this book [142, 156, 164, 165, 180, 181].

Over the past few years, nanolime dispersions have been used for several applications outside of stone and mortar conservation. The successful application of CaLoSiL® in the conservation of paper is described in Refs. [182–185]. Mixing saturated $Ca(OH)_2$ solutions with *iso*-propanol at various temperatures and the subsequent application of the resulting nanolime dispersions for paper conservation is summarised in Ref. [186]. Other papers dealing with the use of nanolime dispersions for paper conservation are Refs. [187, 188].

One special application that was found for a nanolime dispersion in *iso*-propanol is the cleaning and pH adjustment of vegetable-tanned leather [189]. The successful use of CaLoSiL® E25 to eliminate bacteria in human root dentin was described in Ref. [190]

as well as the consolidation of archaeological bone artefacts in [191]. The use of lime for wood conservation has a long tradition. The resulting calcium hydroxide layer has anti-microbial properties, acts a reflective coating, enhances the aesthetic appeal and may prevent insect attack. Similar effects are achieved when nanolime dispersions are used [58b]. An excellent penetration is guaranteed by the small particle size. The alcohol (ethanol or *iso*-propanol) present enhances the anti-microbial properties. The use of nanolime dispersions in the conservation of wooden structures is also described in Ref. [192].

Finally, CaLoSiL® is also applied to modify the susceptibility of materials to microbial growth [193, 194]. The combination of lime and alcohol creates an alkaline milieu and eliminates the basis for microbial growth.

Potentially health damaging effects of nanoparticles are currently an extremely controversially discussed question. All CaLoSiL® products have particle sizes above 50 nm. Recently, a comprehensive study was published concerning this topic. It could be confirmed that nanolime dispersions do not have any cytotoxicity [195].

8.2 Application Recommendations

Any conservation project is a complex task that combines different requirements as well as challenges, mostly from several disciplines. The following section provides an initial overview of the main aspects of the conservation procedure. This is followed by a guideline for the successful application of nanolime dispersions.

The conservation of an object can be divided up into the five steps summarised in Fig. 8.1.

The first part, 'Preliminary investigation of the object', is the basis for all following conservation interventions. The use of non- or minor-destructive characterisation methods is of great importance. A detailed description of the standard analytical methods can be found in section 2.2–2.5. A careful characterisation of the original materials to be treated is an essential pre-condition for the successful application of nanolime dispersions.

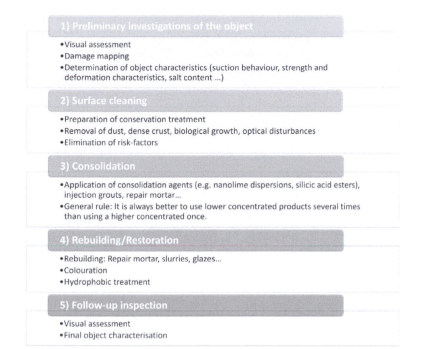

Figure 8.1 Conservation procedure.

Important parameters are:

(1) Overall characteristics of the damage
(2) Mechanical properties of the material in question
(3) Presence of skins (gypsum, alone or in combination with biological growth or crusts formed through air pollution)
(4) Salt and moisture content
(5) Porosity
(6) Suction behaviour
(7) Mineralogical composition
(8) Climatic conditions

In addition to a comprehensive object characterisation, it is always advisable to provide test areas on site and to test nanolime dispersions with different concentrations in different solvents. A detailed description of the entire systematic procedure is provided in Chapter 7 for 13 different historical objects. Beside there are many final exams dealing with nanolime and the consolidation effectiveness for special application areas [196–200].

The main properties of nanolime dispersions, based on laboratory characterisation as well as combination possibilities with other consolidation agents, have been discussed at length in Chapters 4–6. Application recommendations will only be summarised at this point. These are based on our own investigations as well as experience submitted from restorers worldwide. More informations are also available in Ref. [201].

Nanolime dispersions can be applied by all standard techniques such as spraying, immersion or injection (Fig. 8.2). The most suitable application technology has to be chosen on the basis of the characteristics of the substrates and preliminary tests.

All techniques have their own specifics, but it is recommended that the following aspects be taken into account:

- **Spraying** is the most convenient method to consolidate powdery surfaces. The penetration depth is often low, but a significant consolidation of surface layers can be achieved. A very convenient way to treat large areas is to use a **liquid jet pump**. This allows the application of enough fluid for consolidation.
- **Immersion** is commonly used under laboratory conditions.
- In order to fill cracks and detached layers, as well as for the saturation of surfaces with a high absorption capacity, nanolime dispersions can be introduced into the substrates using **syringes** or **pipettes**.
- **Vacuum suction** is always a convenient method to guarantee the complete saturation of porous objects.

Application with a brush or roller is in principle possible, but is not recommended. Brushing is often accompanied by the risk of the surface becoming clogged by fine particles such as dust or the result of biological growth. These are mechanically liberated by the action of the brush. The penetration of nanolime dispersions becomes impossible.

Nanolime dispersions should always be applied at low temperatures (5–25 °C) and under conditions leading to slow evaporation

Figure 8.2 Application possibilities of CaLoSiL®.

rates. A high air humidity (RH > 75%) is advantageous. Moreover, it is important to only treat surfaces that are able to absorb the nanosol. Nanolime dispersions cannot diffuse through dense skins formed by gypsum layers, biological growth or dust. Loose layers beneath dense skins can be consolidated if the nanolime dispersions are introduced into these zones with microdrills. The presence of

Figure 8.3 Karsten tube characterisation of the penetration behaviour of CaLoSiL® E25 in a sandstone sample.

nanolime particles in treated zones can be visualised by spraying 1% phenolphthalein solution in ethanolic solution (70% ethanol, 30% water by volume) onto the treated zones. The appearance of a purple colour indicates alkaline conditions and thus the presence of $Ca(OH)_2$. The precise suction behaviour of substrates can be determined by tests with the Karsten tube. A glass tube thus has to be attached to the surface with plasticine. The tube is filled with the nanolime dispersion. The absorption is determined by measuring the volume absorbed over time (see Section 2.2 and Fig. 8.3). The suction behaviour can be described exactly by plotting the volume of absorbed nanolime dispersion against time. Penetration is often impossible if the materials are wet or water saturated, especially if low porous substrates have to be strengthened. Apart from natural drying, pre-treatment with alcohol will result in the fast removal of water. Nanolime dispersions can be applied after the water-ethanol mixture has evaporated. Transportation is possible only by capillarity in small voids, fissures and cracks. In order to to guarantee deep penetration, it is always necessary to prevent fast blocking of the flow paths. Although the particles are small and the concentration in the standard product is only 25 g calcium hydroxide per litre, near-surface blocking is possible, especially when treating relatively dense materials. In addition, agglomerates formed by the coagulation of nanoparticles due to the presence of high humidity or salts can also lead to blocking. In most cases, zones similar to filter cakes are formed that prevent any further transport of the nanosols into deeper zones. Thus, it is always advisable to begin any treatment with low-concentration nanosols, for example concentrations of only 5 g/L. These can be prepared by diluting, CaLoSiL® E25, for example, with ethanol or *iso*- or *n*-propanol. Materials with a higher concentration can be used in a second step. In the majority of cases, repeated treatment with diluted nanolime dispersions has resulted in better consolidation effects than a single treatment with highly concentrated sols. All CaLoSiL® products can be completely intermixed. An oversaturation of the substrate must be avoided. Any excess nanolime should be sponged off immediately. Figure 8.4 offers a brief overview of the nanolime dispersions that should be used at different object characteristics.

One main problem in many applications is the formation of white haze after nanolime dispersions have been applied. This means that although a good penetration has been achieved, the treated materials are characterised by the appearance of a white-coloured surface after drying, see Fig. 8.5.

Apart from overdosing, the main reason for this is the reverse-migration of the nanoparticles during evaporation of the alcohol. This phenomenon can be observed particularly if the alcohol in the nanolime dispersion evaporates too quickly. What's more, the treatment of humid/wet substrates characterised by a low porosity with relatively high concentrations of nanolime dispersions can lead to the formation of white haze. The following recommendations are made to prevent the formation of white haze:

- Nanolime dispersions should always be applied at low temperatures (5–25 °C) and under conditions leading to slow evaporation rates. The treated surfaces should be protected from rain and direct sunlight for at least 24 h.
- Covering treated surfaces with plastic foil is a good way to prevent rapid evaporation and to guarantee a homogeneous distribution of the nanolime particles in the treated substrates.

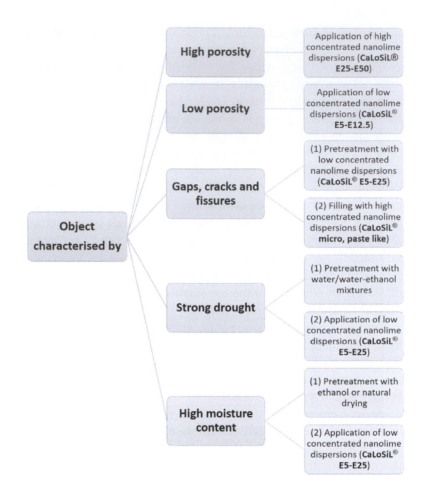

Figure 8.4 Application of nanolime dispersions depending on objective characteristics.

- Favourable curing conditions are also achieved by covering treated surfaces with wet cloths.
- If the characteristics of the substrates (pore radii, suction behaviour, etc.) allow it, small amounts of CaLoSiL® micro should be added to the nanolime dispersion that is used.
- After-treatment with solutions of 0.5 wt% hydroxypropyl cellulose in ethanol-water mixtures (1 : 1 by volume).

Figure 8.5 White haze that fas formed after two treatments with CaLoSiL®
E25.

- After-treatment with water (spraying or application by syringe, pipette).

The after-treatment with water has to be discussed from different points of view:

- All nanolime dispersions form $Ca(OH)_2$ agglomerates with significantly larger particles than in the original product after the addition of water. This means that penetration into narrow structures is limited. On the other hand, if coagulation takes place in the pore space, reverse-migration can be prevented.
- The capillarity of water is much higher than that of ethanol, *iso*-propanol or *n*-propanol. If a capillary active system is treated with nanolime dispersions followed by water, the water pushes the nanolime dispersion deeper into the substrate (Fig. 8.6).
- If a system with narrow capillaries is after-treated with too much water, gel-like calcium hydroxide layers rapidly form on the surface. Enhanced white haze formation is possible.

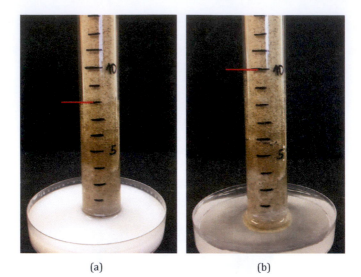

(a)   (b)

Figure 8.6   (a) Capillary suction of CaLoSiL® E25 in fine sand up to a height of 8 cm. (b) Capillary suction of water after treatment with CaLoSiL® leads to an increase of up to 10 cm (red lines indicate the capillary rise).

In conclusion, careful after-treatment with small volumes of water (or water-ethanol mixtures) significantly reduces white haze. However, over-treatment has to be avoided, especially on substrates with a low porosity. In many cases, white haze can already be reduced significantly by the presence of a high relative humidity (80%–90%). Please refer to Chapter 5 and the detailed application reports in Chapter 7 for further discussions of the formation and prevention of white haze.

A significant improvement of the mechanical properties of treated substrates can be achieved by combining **nanolime dispersions and silicic acid esters** (see Chapter 6). Extensive investigations have shown that successive treatment with nanolime and SAEs results in the highest mechanical stabilisation. What is important is that the treatment with nanolime dispersions is realised in the first step. The after-treatment of areas that have initially been treated with SAEs has no positive effects. The nanolime dispersions are unable to penetrate treated zones. They are also unable to convert surfaces with hydrophobic properties into hydrophilic ones. The following recommendations are made for the consecutive application of nanolime dispersions and SAEs:

- In a first step, the substrate has to be saturated with CaLoSiL®. Single or repeated treatments are possible depending on the suction behaviour and the overall conditions. In the latter case, the second and all further treatments should be carried out when the alcohol in the nanolime dispersion has evaporated. Experimental investigations on highly porous mortars have shown that up to six treatments with CaLoSiL® are possible without affecting the penetration behaviour of the SAEs. In most cases, however, a single pre-treatment with the nanolime dispersion is sufficient. The standard product is CaLoSiL® E25, though dispersions with lower or higher nanolime contents may also be expedient.
- The SAE treatment should be realised 24 h after the last application of the nanolime dispersion. Both earlier and later applications have been proven to be connected with a lower final strength of the treated material. The most favourable SAEs are KSE 100 (Remmers, Germany), KSE 300 (Remmers), Protectosil (Evonik, Germany) and Silres BS-OH 100 (Wacker, Germany).

The advantages of a combined, consecutive application of the two stone consolidants are:

- The nanolime dispersions form a layer of fine calcium hydroxide particles on the treated surfaces. These act as an anchor for the SAEs. Consequently, materials that are normally difficult to strengthen with SAEs, mainly calcareous materials, can also be treated.
- The alkaline character of nanolime dispersions results in the accelerated hydrolysis of the SAEs and hydrophilic surfaces are obtained much faster than by SAE treatment only.

The combination of nanolime dispersions and SAEs is particularly advantageous when areas that have no or only limited contact to air have to be treated and carbonation would take a long time. This is the case, when loose zones beneath dense skins have to be consolidated.

Chapter 9

Summary

The conservation and protection of buildings that constitute our cultural heritage is a complex task calling for a comprehensive knowledge of the historical background of the objects as well as the construction technologies and materials used. Whereas the use of organic polymers was encouraged in former decades, today has seen a renaissance of traditional materials such as lime-based mortars, injection grouts or lime suspensions. These are compatible with many historic structures and guarantee long-term stability. One major handicap, however, is that lime suspensions have only a limited penetration capacity. The use of lime water, or more precisely a saturated solution of $Ca(OH)_2$ in water, has no or only an extremely low consolidation capacity. This is due to the low content of dissolved $Ca(OH)_2$ of approximately 1.7 g/L (at 25°C).

One alternative is the use of nanolime dispersions. These are colloidal systems containing nano calcium hydroxide particles in different alcohols. Their synthesis, characterisation and application was the focus of the European project 'STONECORE – Stone Conservation for Refurbishment of Buildings', which was funded within the FP-7 framework under Grant Agreement No. 213651. The results of the research project and the numerous applications (especially those from outside the STONECORE consortium)

Nanomaterials in Architecture and Art Conservation
Gerald Ziegenbalg, Miloš Drdácký, Claudia Dietze, and Dirk Schuch
Copyright © 2018 Pan Stanford Publishing Pte. Ltd.
ISBN 978-981-4800-26-6 (Hardback), 978-0-429-42875-3 (eBook)
www.panstanford.com

show that nanolime dispersions are a promising alternative to conventional stone consolidants. Just like every other conservation agent, their application requires knowledge, skills and experience. The main issues are the formation of white haze as well as the penetration depth. These are easy to overcome, as the results summarised in not only this book but also recent literature show. Another aspect that has only been touched upon as yet concerns the combination of nanolime with traditional conservation agents such as different types of silicic acid esters. The results presented in this book demonstrate the huge potential offered by the combined application of nanolime and SAEs. Further research can be expected to reveal many more conservation opportunities.

The non- or semi-destructive characterisation methods that have been developed allow additional measurements to characterise objects before and after any interventions. One idea of the authors of 'Nanomaterials in Architecture and Art Conservation' was to summarise not only the scientific outcome of the STONECORE project, but also to present their personal experiences, opinions and recommendations. Thus, the documentation of practical applications has taken up a lot of space in this book. Just like all new materials, their comprehensive use will develop over time and requires many successful demonstrations. The applications summarised in this book are only initial examples, many others can be found in literature (see Chapter 8 in particular). Since 2011, nanolime products have been delivered to not only European conservators but also to North America, Asia and the Near East. There has been a sharp rise in demand over the past two years. But of course, one always has to remember that nanolime dispersions are a relatively new consolidation agent. Further research into both fundamental aspects and applications under special conditions is necessary. Possible topics include:

- The combination of nanolime dispersions with different types of silicic acid esters.
- The use of nanolime in the conservation of concrete structures and especially for the realkalisation of cement and treatment of corroded reinforcement.

- The use of nanolime dispersions for paper and wood conservation.
- The development of protective coatings of calcium oxalate, calcium carbonate through the successive treatment of surfaces with nanolime and oxalate solutions as well as oxalic acid esters or carbonic acid esters.

All of the authors of this book are extremely interested in continuing their work on nanolime and are available for discussions on specific subjects as well as for the development of specific projects.

References

1. G. Ziegenbalg, Stone conservation for refurbishment of buidlings (STONECORE): a project funded in the 7th framework programme of the European Union. In M. Krüger, ed. *Cultural Heritage Preservation, Proceedings of the European Workshop on Cultural Heritage Preservation*, Frauenhofer IRBVerlag, 240–245, 2011.

2. ICOMOS, Illustrated glossary on stone deterioration patterns (English-German Version) ICOMOS International scientific committee for stone ISCS, 2010.

3. C. Maierhofer, A. Brink, M. Röllig, and H. Wiggenhauser, Transient thermography for structural investigation of concrete and composites in the near surface region, *Infrared Phys. Technol.*, **43**, 271–278, 2002.

4. G. Wedler, A. Brink, M. Röllig, F. Weritz, and C. Maierhofer, Active infrared thermography in civil engineering – quantitative analysis by numerical simulation, *Federal Institute for Materials Research and Testing (BAM)*, 2003.

5. M. Drdácký and J. Lesák, Non-destructive diagnostics of shallow subsurface defects on masonry. In L. Binda, M. Drdácký, and B. Kasal, eds. *In-Situ Evaluation & Non-Destructive Testing of Historic Wood and Masonry Structures*, 140–147, 2007.

6. P. Castellini, E. Esposito, N. Paone, and E. P. Tomasini, Noninvasive measurements of damage of frescoes paintings and icon by laser scanning vibrometer: experimental results on artificial samples and real works of art, *Measurement*, **28**, 33–45, 2000.

7. J. F. Vignola, Locating faults in wall paintings at the U.S. Capitol by shaker-based laser vibrometry, *APT Bull.*, **36**(1), 25–33, 2005.

8. D. Michoinová, M. Drdácký, and J. Lesák, Inspection and efficiency of consolidation of delaminated parts of historic lime plasters, *Restoration of Architectural Heritage Firenze 2000; CD ROM Proc. Int. Congress CICOP*, Florence, Septempber 17–24, 2000.

9. M. Drdácký and J. Lesák, *Non-Invasive Survey of Detachement of Historic Rendering*, R. Fort, M. Alvarez de Buergo, M. Gomez-Heras, and C. Vazquez-Calvo, eds. Taylor & Francis Group, 2006.

10. R. Sklodowski, M. Drdácký, and M. Sklodowski, Identifying subsurface detachement defects by acoustic tracing, *NDT and E Int.*, **56**, 56–64, 2013.

11. K. J. Beasley, Building facade failures, *Proceedings of the Institutions of Civil Engineers, Forensic Engineering*, **165**, 13–19, 2012.

12. Guide to remedial measures associated with internal plastering, *Mineral Products Association*, 2012.

13. M. Drdácký, Non-standard testing of mechanical characteristics of historic mortars, *Int. J. Archit. Heritage*, **5**, 383–394, 2011.

14. L. Binda, I. Papayianni, E. Toumbakari, and R. Van Hees, Chapter 2.4; Mechanical tests on mortars and assemblages, 1–26, 2004.

15. M. Drdácký and Z. Slížková, Mechanical characteristics of historical mortars from tests on small-sample non-standard specimens, *Mater. Sci. Appl. Chem*, **17**, 21–29, 2008.

16. M. Drdácký, O. Jiroušek, Z. Slížková, J. Valach, and D. Vavrík, Hybrid testing of historic materials, *Experimental Analysis of Nano and Engineering Materials and Structures*, 275–276, 2007.

17. S. Siegesmund and R. Snethlage, *Stone in Architecture*. Springer-Verlag Berlin Heidelberg, 2011.

18. M. Pamplona, M. Kocher, and R. Snethlage, Drilling resistance: overview and outlook, *Z. dt. Ges. Geowiss.*, **158**(3), 665–676, 2007.

19. E. Del Monte and A. Vignoli, In situ mechanical characterization of the mortar in masonry buildings with DRMS, *Proceedings of the RILEM Symposium on Site Assessment of Concrete, Masonry and Timber Structures - SACoMaTiS*, September 1–2, Varenna, 421–430, 2008.

20. M. Drdácký, J. Lesák, K. Niedoba, and J. Valach, Peeling tests for assessing the cohesion and consolidation characteristics of mortar and render surfaces, *Mater. Struct.*, **48**, 1947–1963, 2015.

21. M. Drdácký and Z. Slížková, In situ peeling tests for assessing the cohesion and consolidation characteristics of historic plaster and render surfaces, *Stud. Conserv.*, **60**(2), 121–130, 2015.

22. M. Drdácký, M. Cerný, Z. Slížková, and P. Zíma, Microtube device for innovative water uptake measurements, *Proceedings of the European Workshop on Cultural Heritage Preservation*, 126–130, 2011.

23. J. Valach, J. Bryscejn, M. Drdácký, Z. Slížková, and D. Vavrík, Public perception and optical characterization of degraded historic stone and mortar surfaces, *Proceedings of the International Conference on Heritage, Weathering and Conservation, HWC 2006*, Madrid, **2**, 827–832, 2006.

24. J. Valach and M. Drdácký, An effective method for monitoring and optical characterization of degraded historic stone and mortar surfaces, *Proceedings of the International Workshop*, ICVBC Florence, 37–44, 2008.

25. M. Drdácký, Experimental approach to the analysis of historic timber and masonry structures, *Structural Analysis of Historical Constructions*, New Delhi, **1**, 25–40, 2006.

26. V. Lebrun, C. Toussiant, and E. Pirard, Monitoring color alternation of ornamental flagstones using digital image analysis, *International Conference: Dimension Stone 2004*, Prague, Czech Republic, 139–146, 2004.

27. M. T. Laurenzi and S. Simon, Testing methods and criteria for the selection/evaluation of products for the conservation of porous building materials, *Rev. Conserv.*, **7**, 67–82, 2006.

28. H. R. Sasse and R. Snethlage, Evaluation of stone consolidation treatments, *Sci. Technol. Cult. Heritage*, **5**(1), 85–92, 1996.

29. N. 20/85, Conservazione dei materiali lapidei: manutenzione ordinaria e straordinaria, 1985.

30. F. H. Wittmann and P. Prim, Mesures de l'effet consolidant d'un produit de traitment, *Mater. Constr.*, **16**, 235–242, 1983.

31. A. Bourges, Holistic correlation of physical and mechanical properties of selected natural stones for assessing durability and weathering in the natural environment, Ph.D. dissertation; Fakultät für Geowissenschaften der Ludwigs-Maximilians-Universität München, 201, 2006.

32. T. Frühwirt, *Biaxial flexural strength test device*, TU Bergakademie Freiberg, 2015.

33. M. Drdácký and Z. Slížková, Performance of glauconitic sandstone treated with ethylsilicate consolidation agents, *Proc. 11th Int. Congr. Deterior. Conserv. Stone*, Nicolaus Copernicus University Press, Torun, 1205–1212, 2008.

34. M. Drdácký, D. Frankeová, and Z. Slížková, Ancient sandstone condition assessment in relation to degradation, cleaning and consolidation phenomena, *Geophys. Res. Abstr.*, **17**, 2015.

35. M. Drdácký and Z. Slížková, Lime-water consolidation effects on poor lime mortars. *APT Bull.*, **43**, 31–36, 2012.

36. C. A. Price, The consolidation of limestone using a lime poultice and limewater, *Adhesives and Consolidants*, IIC, London, 160–162, 1984.

37. Z. Slížková and D. Frankeova, Consolidation of porous limestone with nanolime laboratory study, *12th International Congress on the Deterioration and Conservation of Stone Columbia University*, New York, 1–11, 2012.

38. E. Sassoni, G. Graziani, and E. Franzoni, An innovative phosphate-based consolidant for limestone. Part 1: Effectiveness and compatibility in comparison with ethyl silicate, *Constr. Build. Mater.*, **102**, 918–930, 2016.

39. Z. Slížková, M. Drdácký, and A. Viani, Consolidation of weak lime mortars by means of saturated solution of calcium hydroxide or barium hydroxide, *J. Cult. Heritage*, **16**, 452–460, 2014.

40. M. Drdácký, Testing efficiency of stone conservation treatments. In M. Hosseini and I. Karapanagiotis, eds. *Advanced Materials for the Conservation of Stone*. Springer, New York, 175–184, 2018.

41. R. J. Flatt, Predicting salt damage in practice: a theoretical insight into laboratory tests, *RILEM Tech. Lett.*, **2**, 108–118, 2017.

42. A. Arnold and K. Zehnder, Salt weathering on monuments, *1st Int. Symposium on the Conservation of Monuments in the Mediterranean Basin*, Bari, 31–58, 1989.

43. H. J. Schwarz and M. Steiger, Salze und Salzschäden and Bauwerken, *Deutsche Bundesstiftung Umwelt; Workshop im Februar 2008*, Osnabrück, 2008.

44. J. D'ans, D. Bretschneider, H. Eick, and H. E. Freund, Untersuchungen über Calciumsulfate, *Kali und Steinsalz*, **9**, 17–38, 1955.

45. D. E. Garret, *Sodium Sulfate: Handbook of Deposits, Processing, and Use*, Academic Press, 2001.

46. WTA-Merkblatt, 4-5-99/D.

47. B. Middendorf, J. J. Hughes, K. Callebaut, G. Baronio, and I. Papayianni, Investigative methods for the characterisation of historic mortars. Part 2: Chemical characterisation, *Mater. Struct.*, **38**, 771–780, 2005.

48. M. L. Tabasso, Acrylic polymers for the conservation of stone: advantages and drawbacks, *APT Bull.*, **26**, 17–21, 1995.

49. C. Selwitz, *Epoxy Resins in Stone Conservation*, Research in Conservation 7, Getty Conservation Institute, Marina del Rey, CA, 1992.

50. M. Favaro, R. Mendichi, F. Ossola, S. Simon, P. Tomasin, and P. Vigato, Evaluation of polymers for conservation treatments of outdoor exposed stone monuments. Part II: Photo-oxidative and salt-induced weathering of acrylic – silicone mixtures, *Polym. Degrad. Stab.*, **92**, 335–351, 2007.

51. M. Wu, B. Johannesson, and M. Geiker, A review: self-healing in cementitious materials and engineered cementitious composite as a self-healing material, *Constr. Build. Mater.*, **28**, 571–583, 2012.

52. C. Shi and J. Qian, High performance cementing materials from industrial slags: a review, *Resour. Conserv. Recycl.*, **29**, 195–207, 2000.

53. J. Porta, Methodologies for the analysis and characterization of gypsum in soils: a review, *Geoderma*, **87**, 31–46, 1998.

54. E. Charola, J. Pühringer, and M. Steiger, Gypsum: a review of its role in the deterioration of building materials, *Environ. Geol.*, **52**, 339–352, 2007.

55. R. van Hees, R. Veiga, and Z. Slížková, Consolidation of renders and plasters, *Mater. Struct.*, **50**(65), 1–16, 2017.

56. C. Rodriguez-Navarro, E. Hansen, and W. S. Ginell, Calcium hydroxide crystal evolution upon aging of lime putty, *J. Am. Chem. Soc.*, **81**(11), 3032–3034, 1998.

57. M. G. Margalha, A. S. Silva, M. do Rosario Veiga, J. de Brito, R. J. Ball, and G. C. Allen, Microstructural changes of lime putty during aging, *J. Mater. Civ. Eng.*, **25**, 1524–1532, 2013.

58. M. P. Ansell, Multi-functional nano-materials for timber in construction, *Proc. Inst. Civ. Eng. Const. Mater.*, **166**(4), 248–256, 2013.

59. R. Demichelis, P. Raiteri, J. D. Gale, and R. Dovesi, The multiple structures of vaterite, *Cryst. Growth Des.*, **13**, 2247–2251, 2013.

60. L. Kabalah-Amitai, Vaterite crystals contain two interspersed crystal structures, *Science*, **340**, 454–457, 2013.

61. C. Fiori, M. Vandini, S. Prati, and G. Chiavari, Vaterite in the mortars of a mosaic in the Saint Peter basilica, Vatican (Rome), *J. Cult. Heritage*, **10**, 248–257, 2009.

62. H. H. Marey Mahmoud, M. F. Ali, and E. Pavlidou, Characterization of plasters from ptolemaic baths: new excavations near the Karnak temple complex, Upper Egypt, *Archaeometry*, **53**, 693–706, 2011.

63. S. Signorelli, C. Peroni, M. Camaiti, and F. Fratini, The presence of vaterite in bonding mortars of marble inlays from Florence Cathedral, *Mineral. Mag.*, **60**, 663–665, 1996.

64. C. Rodriguez-Navarro, K. Kudłacz, O. Cizer, and E. Ruiz-Agudo, Formation of amorphous calcium carbonate and its transformation into mesostructured calcite, *CrystEngComm*, **17**, 58–72, 2015.

65. L. Goodwin, F. M. Michel, B. L. Phillips, D. A. Keen, M. T. Dove, and R. J. Reeder, Nanoporous structure and medium-range order in synthetic amorphous calcium carbonate, *Chem. Mater.*, **22**, 3197–3205, 2010.

66. C. Rodriguez-Navarro, K. Elert, and R. Ševčík, Amorphous and crystalline calcium carbonate phases during carbonation of nanolimes: implications in heritage conservation, *CrystEngComm*, **18**, 6594–6607 2016.

67. J. D. Rodriguez-Blanco, S. Shaw, and L. G. Benning, The kinetics and mechanisms of amorphous calcium carbonate (ACC) crystallization to calcite via vaterite, *Nanoscale*, **3**, 265–271, 2011.

68. H. Pauly, Ikaite, a new mineral from Greenland, *Arctic*, **16**, 263–264, 1963.

69. L. S. Gomez-Villalba, P. Lopez-Arce, M. Alvarez de Buergo, and R. Fort, Structural stability of a colloidal solution of Ca(OH)2 nanocrystals exposed to high relative humidity conditions, *Appl. Phys. A*, **104**, 1249–1254, 2011.

70. I. Matsushita, Y. Hamada, T. Suzuki, Y. Nomura, T. Moriga, T. Ashida, and I. Nakabayashi, Crystal structure of basic calcium carbonate and its decomposition process in water, *J. Ceram. Soc. Jpn.*, **101**, 1335–1339, 1993.

71. F. Petr, Nekolik kapitol z technologie konzervacních prací, *Zprávy Památkové Péce*, 13–19, 1953.

72. W. Millar and G. Robinson, *Plastering, Plain and Decorative: A Practical Treatise on the Art & Craft of Plastering and Modelling*. London, B.T. Batsford, 1899.

73. S. Peterson, Lime water consolidation. *Proc. Mortars, Cements and Grouts Used in the Conservation of Historic Buildings, ICCROM*, Rome, 53–59, 1982.

74. P. Mora, L. Mora, and P. Philippot, *La conservation des peintures murale*. Bologna, Compositori, 1977.

75. E. Denninger, Die chemischen Vorgänge bei der Festigung von Wandmalereien mit sogenanntem Kalksinterwasser, *Maltechnik*, **64**, 67–69, 1958.

76. N. J. T. Quayale, The case against lime water (or, the futility of consolidating stone with calcium hydroxide), *Conserv. News*, **59**, 68–71, 1996.

77. J. Ashurst and F. G. Dimes, *Conservation of Building and Decorative Stone*. London, Butterworth-Heinemann, 1990.

78. Ashurst and N. Ashurst, *Practical Building Conservation*, Vol. 1 Stone Masonry. Aldershot, Gower, 1988.

79. J. Ashurst and N. Ashurst, *Practical Building Conservation*, Vol. 3 Mortars, Plasters and Renders. Aldershot, Gower, 1988.

80. M. Drdácký and Z. Slížková, Calcium hydroxide based consolidation of lime mortars and stone, Institute of Theoretical and Applied Mechanics of the Academy of Sciences of the CZ, *Tech. Rep.*, 2008.

81. C. Price, K. Ross, and G. White, A further appraisal of the 'lime technique' for limestone consolidation, using a radioactive tracer, *Stud. Conserv.*, **33**, 178–186, 1988.

82. I. Brajer and N. Kalsbeek, Limewater absorption and calcite crystal formation on a limewater-impregnated secco wall painting, *Stud. Conserv.*, **44**, 145–156, 1999.

83. M. Tavares, M. do Rosário Veiga, and A. Fragata, Conservation of old renderings - the consolidation of rendering with loss of cohesion, *Historical Mortars Conference*, *HMC*, Lisabon, 2008.

84. J. Fidler, Lime treatments: an overview of lime watering and shelter coating of friable historical limestone masonries. In J. Fidler, ed. *Stone: Stone Building Materials, Construction and Associated Component Systems; Their Decay and Treatment*, English Heritage Research Transactions 2. London, James & James, 19–28, 2002.

85. C. Woolfit and N. Durnan, Lime method evaluation: a survey of sites where lime-based conservation techniques have been employed to treat decaying limestones. In J. Fidler, ed. *Stone: Stone Building Materials, Construction and Associated Component Systems; Their Decay and Treatment*, English Heritage Research Transactions 2. London, James & James, 29–44, 2002.

86. E. Hansen, E. Doehne, J. Fidler, J. Larson, B. Martin, M. Matteini, C. Rodriguez-Navarro, E. S. Pardo, C. Price, A. de Tagle, J. M. Teutonico, and N. Weiss, A review of selected inorganic consolidats and protective treatments for porous calcareous meterials, *Stud. Conserv.*, **48**, 13–25, 2003.

87. E. F. Doehne and C. A. Price, *Stone Conservation: An Overview of Current Research*, 2nd ed. Los Angeles, The Getty Conservation Institute, 2010.

88. M. Matteini, S. Rescic, F. Fratini, and G. Botticelli, Ammonium phosphates as consolidating agents for carbonatic stone materials

used in architecture and cultural heritage: preliminary research, *Int. J. Archit. Heritage*, **5**, 717–736, 2011.

89. P. Baglioni, D. Chelazzi, and R. Giorgi, *Nanotechnologies in the Conservation of Cultural Heritage*. Springer, 2015.

90. P. Baglioni and R. Giorgi, Soft and hard nanomaterial for restoration and conservation of cultural heritage, *Soft Matter*, **2**, 293–303, 2006.

91. A. Thorn, Lithium silicate consolidation of wet stone and plaster, *12th International Congress on the Deterioration and Conservation of Stone Columbia University*, New York, 2012.

92. M. H. Oh, J. H. So, J. D. Lee, and S. M. Yang, Preparation of silica dispersion and its phase stability in the presence of salts, *Korean J. Chem. Eng.*, **16**(4), 532–537, 1999.

93. R. Zarraga, D. E. Alvarez-Gasca, and J. Cervantes, Solvent effect on TEOS film formation in the sandstone consolidation process, *Silicon Chem.*, **1**, 397–402, 2002.

94. E. Sassoni, S. Naidu, and G. W. Scherer, The use of hydroxyapatite as a new inorganic consolidant for damaged carbonate stones, *J. Cult. Heritage*, **12**, 346–355, 2011.

95. E. Franzoni, E. Sassoni, and G. Graziani, Brushing, poultice or immersion? The role of the application technique on the performance of a novel hydroxyapatite-based consolidating treatment for limestone, *J. Cult. Heritage*, **16**, 173–184, 2015.

96. A. Górniak, The use of hydroxy-apatite for consolidation of calcareous stones: light limestone pinczow and gotland sandstone (Part I), *Proc. 13th Int. Congr. Deterioration Conserv. Stone, University of the West of Scotland*, 2016.

97. G. Graziani, E. Sassoni, E. Franzoni, and G. W. Scherer, Marble protection by hydroxyapatite coatings, *Proc. 13th Int. Congr. Deterioration Conserv. Stone, University of the West of Scotland*, 2016.

98. G. W. Scherer and E. Sassoni, Mineral consolidants, *Int. RILEM Conf. Mater. Syst. Struct. Civ. Eng.*, 2016.

99. Liu, B. Zhang, Z. Shen, and H. Lu, A crude protective film on historic stones and its artificial preparation through biomimetic synthesis, *Appl. Surf. Sci.*, **253**, 2625–2632, 2006.

100. B. Doherty, M. Pamplona, C. Miliani, M. Matteini, A. Sgamellotti, and B. Brunetti, Durability of the artificial calcium oxalate protective on two Florentine monuments, *J. Cult. Heritage*, **8**, 186–190, 2007.

101. E. Boquet, A. Boronat, and A. Ramos-Cormenzana, Production of calcite (calcium carbonate) crystals by soil bacteria is a general phenomenon, *Nature*, **246**, 527–529, 1973.

102. J. M. Jakutaj, J. W. Lukaszewicz, and J. Karbowska-Berent, Study of consolidation of porous and dense limestones by bacillus cereus biomineralization, *13th International Congress on the Deterioration and Conservation of Stone, University of the West of Scotland*, 2016.

103. M. T. Gonzalez-Munoz, C. Rodriguez-Navarro, C. Jimenez-Lopez, and M. Rodriguez-Gallego, Method and product for protecting and reinforcing construction and ornamental materials, Patent WO 2008/009 771 A1, 2008.

104. C. Rodriguez-Navarro, F. Jroundi, and M. T. Gonzalez-Muñoz, Stone consolidation by bacterial carbonatogenesis: evaluation of in situ applications, *Restor. Build. Monum.*, **21**, 9–20, 2015.

105. W. de Muynck, N. de Belie, and W. Verstraete, Microbial carbonate precipitation in construction materials: a review, *Ecol. Eng.*, **36**, 118–136, 2010.

106. W. de Muynck, K. Cox, N. de Belie, and W. Verstraete, Bacterial carbonate precipitation as an alternative surface treatment for concrete, *Constr. Build. Mater.*, **22**, 875–885, 2008.

107. A. Sierra-Fernandez, L. S. Gomez-Villalba, M. E. Rabanal, and R. Fort, New nanomaterials for applications in conservation and restoration of stony materials: a review, *Materiales de Construccion*, **67**, 1–18, 2017.

108. M. Ambrosi, L. Dei, R. Giorgi, C. Neto, and P. Baglioni, Colloidal particles of Ca(OH)2: properties and applications to restoration of frescoes, *Langmuir*, **17**, 4251–4255, 2001.

109. B. Salvadori and L. Dei, Synthesis of Ca(OH)2 nanoparticles from diols, *Langmuir*, **17**(8), 2371–2374, 2001.

110. B. Delfort, M. Born, A. Chivé, and L. Barré, Colloidal calcium hydroxide in organic medium: synthesis and analysis, *J. Colloid Interface Sci.*, **189**, 151–157, 1997.

111. A. Nanni and L. Dei, Ca(OH)2 nanoparticles from W/O microemulsions, *Langmuir*, **19**, 933–938, 2003.

112. C. Rodriguez-Navarro, A. Suzuki, and E. Ruiz-Agudo, Alcohol dispersions of calcium hydroxide nanoparticles for stone conservation, *Langumir*, **29**(11), 457–470, 2013.

113. G. Taglieri, V. Daniele, G. D. Re, and R. Volpe, A new and original method to produce Ca(OH)2 nanoparticles by using an anion exchange resin, *Adv. Nanopart.*, **4**, 17–24, 2015.

114. T. Liu, Y. Zhu, X. Zhang, T. Zhang, T. Zhang, and X. Li, Synthesis and characterization of calcium hydroxide nanoparticles by hydrogen plasma-metal reaction method, *Mater. Lett.*, **64**, 2575–2577, 2012.

115. B. Gonzalez, N. Calvar, E. Gomez, and A. Dominguez, Density, dynamic viscosity, and derived properties of binary mixtures of methanol or ethanol with water, ethyl acetate, and methyl acetate at T = (293.15, 298.15, and 303.15) K, *J. Chem. Thermodyn.*, **39**, 1578–1588, 2007.

116. G. Akerlof, Dielectric constants of some organic solvent-water mixtures at various temperatures, *J. Am. Chem. Soc.*, **54**, 4125–4139, 1932.

117. F. M. Pang, E. E. Seng, T. T. Teng, and M. H. Ibrahim, Densities and viscosities of aqueous solutions of 1-propanol and 2-propanol at temperatures from 293.15 K to 333.15 K, *J. Mol. Liq.*, **136**, 71–78, 2007.

118. P. D'Armada and E. Hirst, Nano-lime for consolidation of plaster and stone, *J. Archit. Conserv.*, **18**(1), 63–80, 2012.

119. K. S. Keller, M. H. M. Olsson, M. Yang, and S. L. S. Stipp, Adsorption of ethanol and water on calcite: dependence on surface geometry and effect on surfcae behavior, *Langmuir*, **31**, 3847–3853, 2015.

120. D. Konopacka-Lyskawa and M. Lackowski, Influence of ethylene glycol on $CaCO3$ particles formation via carbonation in the gasslurry system, *J. Cryst. Growth*, **321**, 136–141, 2011.

121. T. Waly, M. D. Kennedy, G. J. Witkamp, G. Amy, and J. C. Schippers, The role of inorganic ions in the calcium carbonate scaling of seawater reverse osmosis systems, *Desalination*, **284**, 279–287, 2012.

122. G. L. Pesce, Study of carbonation in novel lime based material, Thesis; University of Bath, Department of Architecture and Civil Engineering, 2014.

123. S. Ali and A. Ali S, Determination of thermodynamic parameters from the dissolution of calcium hydroxide in mixed solvent systems by pH-metric method, *J. Phys. Chem. Biophys.*, **3**(2), 1–6, 2013.

124. O. Cizer, C. Rodriguez-Navarro, E. Ruiz-Agudo, J. Elsen, D. van Gemert, and K. van Balen, Phase and morphology evolution of calcium carbonate precipitated by carbonation of hydrated lime, *J. Mater. Sci.*, **47**, 6151–6165, 2012.

125. T. Beruto and R. Botter, Liquid-like H2O adsorption layers to catalyze the Ca(OH)2/CO2 solid-gas reaction and to form a non-protective solid product layer at 20°C, *J. Eur. Ceram. Soc.*, **20**, 497–503, 2000.

126. G. Montes-Hernandez, A. Pommerol, F. Renard, P. Beck, E. Quirico, and O. Brissaud, In situ kinetic measurements of gas-solid carbonation of

Ca(OH)2 by using an infrared microscope coupled to a reaction cell, *Chem. Eng. J.*, **161**, 250–256, 2010.

127. E. Ruiz-Agudo, K. Kudacz, C. V. Putnis, A. Putnis, and C. Rodriguez-Navarro, Dissolution and carbonation of portlandite [Ca(OH)2] single crystals, *Environ. Sci. Technol.*, **47**(11), 342–349, 2013.

128. Dalmolin, E. Skovroinski, A. Biasi, M. L. Corazza, C. Dariva, and V. Oliveira, Solubility of carbon dioxide in binary and ternary mixtures with ethanol and water, *Fluid Phase Equilib.*, **245**, 193–200, 2006.

129. C. Rodriguez-Navarro, I. Vettori, and E. Ruiz-Agudo, Kinetics and mechanism of calcium hydroxide conversion into calcium alkoxides: implications in heritage conservation using nanolimes, *Langmuir*, **32**, 5183–5194, 2016.

130. S. F. Chen, H. Cölfen, M. Antonietti, and S. H. Yu, Ethanol assisted synthesis of pure and stable amorphous calcium carbonate nanoparticles, *Chem. Commun.*, **49**, 9564–9566, 2013.

131. L. S. Gomez-Villalba, P. Lopez-Arce, and R. Fort, Nucleation of CaCO3 polymorphs from a colloidal alcoholic solution of Ca(OH)2 nanocrystals exposed to low humidity conditions, *Appl. Phys. A*, **106**, 213–217, 2012.

132. K. Sand, J. D. Rodriguez-Blanco, E. Makovicky, L. G. Benning, and S. L. S. Stipp, Crystallisation of CaCO3 in water - alcohol mixtures: spherulitic growth, polymorph stabilization, and morphology change, *Cryst. Growth Des.*, **12**, 842–853, 2012.

133. A. Daehne and C. Herm, Calcium hydroxide nanosols for the consolidation of porous building materials - results from EUSTONECORE, *Heritage Sci.*, **1**, 1–9, 2013.

134. K. Schönhofer, Calciumhydroxid-Sol–Ein Festigungsmittel für historische Putze, Diplomarbeit, 2006.

135. A. Soleimani Dorcheh and M. H. Abbasi, Silica aerogel; synthesis, properties and characterization, *J. Mater. Proc. Technol.*, **199**, 10–26, 2008.

136. G. Wheeler, *Alkoxysilanes and the Consolidation of Stone.* Los Angeles, The Getty Conservation Institute, 2005.

137. G. Khaskin, Several applications of deuterium and heavy oxygen in the chemistry of flint, *Dokl. Akad. Nauk SSSR*, **85**, 129, 1952.

138. J. Brinker and G. W. Sherer, *The Physics and Chemistry of Sol-Gel Processing.* Academic Press, New York, 1990.

139. R. K. Iler, *The Chemistry of Silica: Solubility, Polymerization, Colloid and Surface Properties and Biochemistry of Silica*. Wiley, New York, 1979.

140. D. Keefer, The effect of hydrolysis conditions on the structure and growth of silicate polymers, *MRS Proc.*, **32**, 15–24, 2011.

141. S. M. Briffa, E. Sinagra, and D. Vella, TEOS based consolidants for maltese globigerina limestone: effect of hydroxyl conversion treatment, *12th International Congress on the Deterioration and Conservation of Stone Columbia University*, New York, **12**, 2012.

142. G. Ziegenbalg and E. Piasczcynski, The combined application of calcium hydroxide nano-sols and silicic acid ester – a promising way to consolidate stone and mortar, *Proceeding 12th Int. Congr. Deterioration Conserv. Stone*, New York, 2012.

143. D. Macounová, Restoration of an angel limestone statue from the House No. 48 in Kutná Hora using nanosols on a calcium hydroxide base for the consolidation of organodetritic limestone, *Extended Restoration Report*, 218, 2011.

144. J. Doubal, Kamenné památky Kutné Hory: restaurování a péce o socharská díla, Pardubice, 2015.

145. K. Bayer and L. Machacko, Calcium hydroxide nanosuspensions as consolidants of porous limestone and lime plasters - from laboratory tests to practical application, Conference Paper, 2012.

146. L. Machacko, Exploration and restoration of historical plasters on the 2nd floor of the cloister of the former monastery Rosa Coeli in Dolní Kounice, Restoration Report, University of Pardubice, Faculty of Restoration, 2011.

147. K. Bayer, Chemical-technological exploration of plasters at the cloister of the former monastery Rosa Coeli in Dolní Kounice. In M. Luboš, et. al. *Exploration and Restoration of Historical Plasters on the 2nd Floor Cloister of the Former Monastery Rosa Coeli in Dolní Kounice, 2011*. p. 6. Restoration report. University of Pardubice, Faculty of Restoration, 2009.

148. J. Koblischek, The consolidation of natural stone with a stone strengthener on the basis of poly-silicic-ethylester, *Proc. 8th Int. Congr. Deterioration Conserv. Stone*, 1191, 1996.

149. J. Válek and J. J. Hughes, eds. Nano-lime as a binder for injection grouts and repair mortars, *Historic Mortars - HMC 2010 and RILEM TC 203-RHM final workshop*, 2010.

150. M. Drdácký, J. Lesák, S. Resic, Z. Slížková, P. Tiano, and J. Valach, Standardization of peeling tests for assessing the cohesion and

consolidation characteristics of historic stone surfaces, *Mater. Struct.*, **45**, 505–520, 2012.

151. M. Drdácký, Z. Slížková, and J. Valach, Príspevek technických ved k záchrane a restaurování památek, 171–186, 2015.

152. R. Giorgi, L. Dei, and P. Baglioni, A new method for consolidating wall paintings based on dispersions of lime in alcohol, *Stud. Conserv.*, **45**, 154–161, 2000.

153. J. Vojtechovsky, Surface consolidation of wall paintings using lime nano-suspensions, *Acta Polytech.*, **57**(2), 139–148, 2017.

154. I. Kociánová, Restaurování centrálního výjevu na klenbe kaple sv. Isidora v Krenove, Master Thesis; University of Pardubice, Faculty of Restoration, 2013.

155. L. Slouková, Restaurování výjevu saoznacením "Archangeli" na klenbe kaple sv. Isidora v Krenove, Master Thesis; University of Pardubice, Faculty of Restoration, 2015.

156. M. Piaszczynski, V. Wolf, and E. Ghaffari, The combination of calcium hydroxide-sol and silicic acid ester as new method for the structural consolidation of objects built of tuff, lime marl, trachyte-latest findings, *12th International Congress on the Deterioration and Conservation of Stone Columbia University*, New York, 2012.

157. E. Ghaffari, T. Köberle, and J. Weber, Methods of polarizing microscopy and SEM to assess the performance of nano-lime consolidants in porous solids, *12th International Congress on the Deterioration and Conservation of Stone Columbia University*, New York, 2012.

158. A. Zornoza-Indart, P. Lopez-Arce, N. Leal, J. Simao, and K. Zoghlami, Consolidation of a Tunisian bioclastic calcarenite: from conventional ethyl silicate products to nanostructured and nanoparticle based consolidants, *Constr. Build. Mater.*, **116**, 188–202, 2016.

159. A. Bonazza, G. Vidorni, I. Natali, C. Giosuè, F. Tittarelli, and C. Sabbioni, Field exposure tests to evaluate the efficiency of nano-structured sonsolidants on carrara marble, *13th International Congress on the Deterioration and Conservation of Stone, University of the West of Scotland*, 2016.

160. L. Dei and B. Salvadori, Nanotechnology in cultural heritage conservation: nanometric slaked lime saves architectonic and artistic surfaces from decay, *J. Cult. Heritage*, **7**, 110–115, 2006.

161. J. Otero, A. E. Charola, C. A. Grissom, and V. Starinieri, An overview of nanolime as a consolidation method for calcareous substrates, *Geconservación*, **1**(11), 71–78, 2017.

162. G. Ziegenbalg, Colloidal calcium hydroxide - a new material for consolidation and conservation of carbonatic stones, *11th International Congress on Deterioration and Conservation of Stone*, Torun, 1109–1115, 2008.

163. G. Ziegenbalg, K. Bruemmer, and J. Pianski, Nano-lime - a new material for the consolidation and conservation of historic mortars, *2nd Historic Mortars Conference HMC2010 and RILEM TC 203-RHM Final Workshop, 22–24 September 2010*, Prague, Czech Republic, 1301–1309, 2010.

164. G. Ziegenbalg and M. Dobrzynska-Musiela, Nanolime- possible applications for the conservation and protection of the cultural heritage, *Rehabend 2016, 24–27 May*, Burgos, Spain, 2016.

165. E. Piaszcynski, P. Egloffstein, and G. Ziegenbalg, The portal in Tholey-unconventional method for the preservation of scaling and shelled sandstone, *11th International Congress on Deterioration and Conservation of Stone*, Torun, 2008.

166. B. Newman, A nanolime case study: the City of London Cemetery entrance screen, *BCD Special Report on Historic Churches 24th Annual Edition*.

167. A. Attwood and K. Reczek, Sgraffito conservation at the Henry Cole wing of the Victoria and Albert museum, *ASCHB Transactions*, **38**, 2015.

168. M. Lanzón, J. A. Madrid, A. Martínez-Arredondo, and S. Mónaco, Use of diluted Ca(OH)2 suspensions and their transformation into nanostructured CaCO3 coatings: a case study in strengthening heritage materials (stucco, adobe and stone), *Appl. Surf. Sci.*, 1–8, 2017.

169. Z. Slížková, K. Bayer, and J. Weber, Konservierung von Leithakalken auf Basis von Calciumhydroxid-Nanopartikeln, *Nanolith*, 2016.

170. C. Rodriguez-Navarro and E. Ruiz-Agudo, Nanolimes: from synthesis to application, *Pure Appl. Chem.*, 2–28, 2017.

171. M. L. Abd el Latif and A. A. Shakal, The nanolime as a consolidating material for the ornamental hardened limestone of Sheikh Fadl, *Int. J. Sci. Eng. Res.*, **7**, 39–50, 2016.

172. G. L. Pesca, D. Morgan, D. Odgers, A. Henry, M. Allen, and J. R. Ball, Consolidation of weathered limestone using nanolime, *Proc. Inst. Civ. Eng. Const. Mater.*, **166**, 213–228, 2013.

173. P. López-Arce, L. S. Gómez-Villalba, S. Martínez-Ramírez, M. Álvarez de Buergo, and R. Fort, Influence of relative humidity on the carbonation of calcium hydroxide nanoparticles and the formation of calcium carbonate polymorphs, *Powder Technol.*, **205**, 263–269, 2011.

174. G. Borsoi, B. Lubelli, R. van Hees, R. Veiga, and A. Silva, Understanding the transport of nanolime consolidants within Maastricht limestone, *J. Cult. Heritage*, **18**, 242–249, 2016.

175. G. Borsoi, B. Lubelli, R. van Hees, R. Veiga, and A. Silva, Optimization of nanolime solvent for the consolidation of coarse porous limestone, *Appl. Phys. A*, **122**, 1–10, 2016.

176. G. Borsoi, B. Lubelli, R. van Hees, R. Veiga, A. Santos Silva, L. Colla, L. Fedele, and P. Tomasin, Effect of solvent on nanolime transport within limestone: how to improve in-depth deposition, *Colloids Surf. A: Physicochem. Eng. Aspects*, **497**, 171–181, 2016.

177. K. Niedoba, Z. Slížková, D. Frankeova, C. L. Nunes, and I. Jandesjsek, Modifying the consolidation depth of nanolime on Maastrich limestone, *Constr. Build. Mater.*, **133**, 51–56, 2017.

178. A. Arizzi, L. S. Gomez-Villalba, P. Lopez-Arce, G. Cultrone, and R. Fort, Lime mortar consolidation with nanostructured calcium hydroxide dispersions: the efficacy of different consolidating products for heritage conservation, *Eur. J. Mineral.*, **27**, 1–13, 2015.

179. P. Lopez-Arce and A. Zornoza-Indart, Carbonation acceleration of calcium hydroxide nanoparticles induced by yeast fermentation, *Appl. Phys. A*, **120**, 1475–1495, 2015.

180. B. Frohberg, Steinkonservierung: Der Dom zu Xanten, *Restauro*, 22–27, 2016.

181. G. Borsoi, R. Veiga, and A. S. Silva, Effect of nanostructured lime-based and silica-based products on the consolidation of historical renders, *3rd Historic Mortars Conference; 11–14 September 2013*, Glasgow, Scotland, 2013.

182. A. Michalcová, O. Nedela, T. Tribulová, M. Durovic, and M. Šlouf, Utilization of Ca(OH)2 nanoparticles in paper deacidification, *4th International Conference Nanocon Czech Republic*, 2012.

183. G. Poggi, N. Toccafondi, L. N. Melita, J. C. Knowles, L. Bozec, R. Giorgi, and P. Baglioni, Calcium hydroxide nanoparticles for the conservation of cultural heritage: new formulations for the deacidification of cellulose-based artifacts, *Appl. Phys. A*, **114**, 685–693, 2014.

184. S. Sequeira, C. Casanova, and E. J. Cabrita, Deacidification of paper using dispersions of Ca(OH)2 nanoparticles in isopropanol. Study of efficiency, *J. Cult. Heritage*, **7**, 264–272, 2006.

185. G. Poggi, R. Giorgi, N. Toccafondi, V. Katzur, and P. Baglioni, Hydroxide nanoparticles for deacidification and concomitant inhibition of iron-gall ink corrosion of paper, *Langmuir*, **26**(24), 19084–19090, 2010.

186. S. Bastone, D. F. Chillura Martino, V. Renda, M. L. Saladino, G. Poggi, and E. Caponetti, Alcoholic nanolime dispersions obtained by the insolubilisation - precipitation method and its application for the deacidification of ancient paper, *Colloids Surf. A: Physicochem. Eng. Aspect*, **513**, 241–249, 2017.

187. G. Poggi, N. Toccafondi, D. Chelazzi, P. Canton, R. Giorgi, and P. Baglioni, Calcium hydroxide nanoparticles from solvothermal reaction for the deacidification of degraded waterlogged wood, *J. Colloid Interface Sci.*, **473**, 1–8, 2016.

188. E. Stefanis and C. Panayiotou, Protection of lignocellulosic and cellulosic paper by deacidification with dispersions of micro- and nano-particles of Ca(OH)2 and Mg(OH)2 in alcohols, *Restaurator*, **28**, 185–200, 2007.

189. P. Baglioni, A. Bartoletti, L. Bozec, D. Chelazzi, R. Giorgi, M. Odlyha, D. Pianorsi, G. Poggi, and P. Baglioni, Nanomaterials for the cleaning and pH adjustment of vegetable-tanned leather, *Appl. Phys. A*, **122**, 114, 2016.

190. P. Louwakul, A. Saelo, and S. Khemaleelakul, Efficacy of calcium oxide and calcium hydroxide nanoparticles on the elimination of Enterococcus faecalis in human root dentin, *Clin. Oral Invest.*, **21**, 865–871, 2016.

191. A. Palazzo, B. Megna, I. Reiche, and J. Levy, Comparative study between four consolidation systems suitable for archaeological bone artefacts, 103–108, 2015.

192. R. Giorgi, D. Chelazzi, and P. Baglioni, Nanoparticles of calcium hydroxide for wood conservation. The deacidification of the vasa warship, *Langmuir*, **21**(10), 743–748, 2005.

193. G. Ziegenbalg and P. D. Askew, The use of novel, nano-lime dispersions to modify the susceptibility of materials to microbial growth, *Biocides in Synthetic Materials*, Berlin, Germany, 1–8, 2010.

194. P. Askew, Fungal and algal growth on stone and the use of calcium hydroxide nano-dispersion-based consolidants on its remediation, *Stone Conservation, Restoration and Assessment of Monuments and Objects of Art, Interim Report STONECORE Project*, 2011.

195. E. Tedesco, I. Micetic, S. G. Ciappellano, C. Micheletti, M. Venturini, and F. Benetti, Cytotoxicity and antibacterial activity of a new generation of nanoparticle-based consolidants for restoration and contribution to the safe-by-design implementation, *Toxicol. in Vitro*, **29**, 1736–1744, 2015.

196. G. Borsoi, Nanostructured lime-based materials for the conservation of calcareous substrates, Ph.D. dissertation; Delft University of Technology, Faculty of Architecture and the Built Environment, Chair of Heritage & Technology, 2017.

197. L. Hull, Can nanolime stone consolidation offer a feasible conservation method for limestone ecclesiastical buildings? Ph.D. dissertation; University of the West of England, Bristol, 2012.

198. D. E. Lohmas, Evaluating the effectiveness of calcium hydroxide nanoparticle dispersions for the consolidation of painted earthen architectural surfaces, Ph.D. dissertation; University of California Los Angeles, 2012.

199. A. S. Campell, Consolidant particle transport in limestone, concrete and bone, Ph.D. dissertation; The University of Edinburgh; Institute for Materials and Processes School of Engineering, 2013.

200. J. Dunajská, Tests on sustainability of consolidation treatments with CaLoSil nanosuspensions on plaster reference samples, Ph.D. dissertation; University of Pardubice; Faculty of Restoration, 2012.

201. D. Odgers, *Nanolime: A Practical Guide to its Use for Consolidating Weathered Limestone*, Historic England, 2017.

Index

T - #0108 - 291019 - C476 - 229/152/21 [23] - CB - 9789814800266